THE TRAVEL DIARIES OF ALBERT EINSTEIN

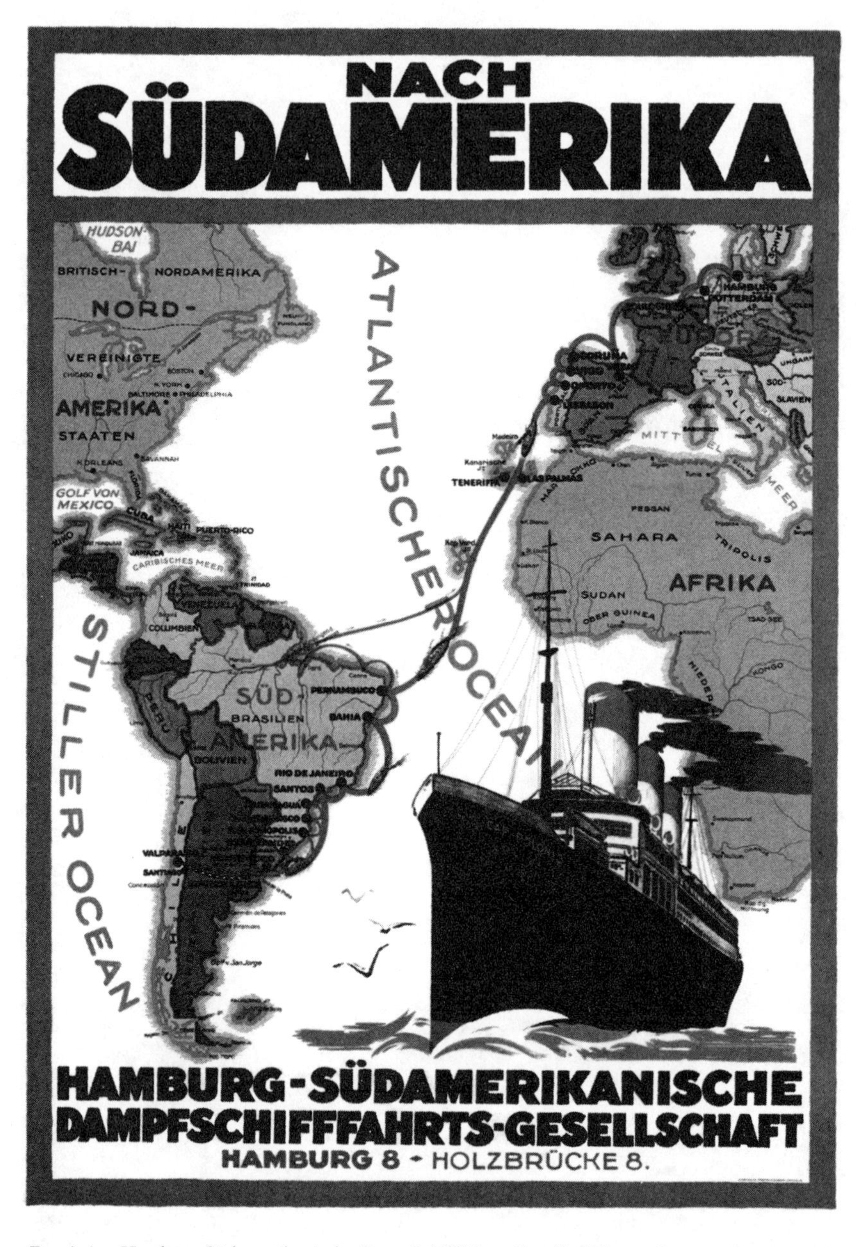

Frontispiece. Hamburg-Südamerikanische Dampfschifffahrts-Gesellschaft travel poster showing Einstein's route to South America on the S.S. *Cap Polonio* (Artwork by Ottomar Anton).

THE TRAVEL DIARIES OF

Albert Einstein

EDITED BY

ZE'EV ROSENKRANZ

SOUTH AMERICA

1925

Albert Einstein.

PRINCETON UNIVERSITY PRESS

PRINCETON AND OXFORD

Published by Princeton University Press
41 William Street, Princeton, New Jersey 08540
99 Banbury Road, Oxford OX2 6JX

press.princeton.edu

ISBN 9780691201023
ISBN (e-book) 9780691242507

Library of Congress Control Number 2022002226

British Library Cataloging-in-Publication Data is available

Editorial: Eric Crahan, Barbara Shi
Production Editorial: Terri O'Prey
Cover and Text Design: Chris Ferrante
Production: Danielle Amatucci
Publicity: Sara Henning-Stout, Kate Farquhar-Thomson

Cover images: (front, top left) Einstein in Buenos Aires, late March 1925. Courtesy Archivo General de la Nación, Buenos Aires. (front, top right) Einstein's South America travel diary. Courtesy of the Albert Einstein Archives. (front, bottom) S.S. *Cap Polonio* in Rio de Janeiro. BTEU / Gerfototek / Alamy Stock Photo. (back) Nach Südamerika poster. Photo by Swim Ink 2, LLC / Corbis / Getty Images.

This book has been composed in Kis Antiqua Now

Printed on acid-free paper. ∞

Printed in the United States of America

10 9 8 7 6 5 4 3 2 1

To Vanessa
and her wonderful South American family

Contents

Illustrations

Preface

"THE DAMN FLU IS ON THE LOOSE AGAIN; it's raging eerily here as well."[1] Thus wrote Albert Einstein in October 1918 from Berlin to his elder son Hans Albert who was living in Zurich. The influenza pandemic had broken out in the summer and was taking its toll on local populaces in Germany and Switzerland. Both his mother Pauline in Lucerne and his future step-daughter Margot in Berlin had recently fallen ill with the disease but would fortunately survive.[2] Einstein decided not to travel to Switzerland due to the pandemic, among other reasons, even though he had not seen his young sons in over a year.[3] Around the time of the armistice that would end World War I he wrote to his estranged wife Mileva: "Here also the flu is very potent and virulent; I have been spared until now, though."[4] Einstein himself would not contract the illness.

When I started working on this edition in 2019, the 1918–1919 pandemic was merely an historical event devoid of much obvious personal relevance. The only social distancing I had ever heard of was my late mother's account of cinema audiences sitting apart from each other during polio epidemics in Australia in the 1940s and 1950s. Like many others, I obliviously never imagined I would experience anything similar. I knew about Einstein's comments on the pandemic he had lived through a century before and could even allow myself to feel judgmental about his possible lack of courage in refraining from visiting his sons in Switzerland. Two years on and having now gone through a pandemic myself and having also made difficult decisions about travel in a period of a global outbreak, I can certainly empathize with his dilemma far more easily. By the time Einstein embarked on his three-month voyage to South America in the spring of 1925, the deadly influenza of six years earlier had lost much of its virulence. This was his third trip overseas after travels to the United States and the Far East. As we will see, in contrast to those journeys, this was not one he looked forward to but felt compelled to embark on. That makes the trip possibly all the more fascinating.

Working on this volume has made me reflect on my earliest memories relating to South America. I was a small child when my parents took me to the local cinema in Melbourne and we saw a short film about a young boy who dreams of going to Maracanã stadium in Rio de Janeiro to see Pelé play, which must have been from around 1970 when Brazil won the World Cup. Growing up, the closest I had ever come to visiting South America was passing through the Panama Canal where I was fascinated by the transcontinental Bridge of the Americas at the entrance to the canal and briefly stopped over on the island of Curaçao just off the coast of Venezuela. A few years later I saw the Brazilian movie *Macunaíma* on Austrian television as an adolescent in Vienna, not comprehending much of the film's magic realism but developing a crush on its female star Dina Sfat. Learning of the brutal excesses against *los desaparecidos* committed by the military dictatorships in Argentina and Chile also formed part of my burgeoning political awareness around that period.

Soon after I started working at the Albert Einstein Archives at the Hebrew University in Jerusalem in the late 1980s, I perused his travel diaries, among them this one on South America. I felt a special affinity to these journals, having traveled between continents myself by sea in my preadolescent years. They became my favorite Einstein documents and they still are. In carrying out the research for this edition, I have had the opportunity to reflect on my own international journeys, humbly comparing my own experiences to those of a global icon. What I enjoy the most about his travel diaries are their authenticity. In them, Einstein expresses his innermost, immediate thoughts. He does not seem to filter his entries, not even for the sake of the very limited audience of family members waiting to read about his adventures back in Berlin. These very personal documents reveal Einstein in his most unadulterated form. They offer an insight into his most uncensored impressions and reflections. The sight is not always a pretty one. We gain a glimpse not only into what enchants and intrigues him but also into what he abhors, disdains, and views with intolerance. This provides us with a remarkable opportunity to examine our own prejudices and biases. I therefore feel privileged and delighted to share this fascinating record of Einstein's journey to far-off lands with a wider audience.

Pasadena, California, and São Paulo, Brazil, May 2021

Acknowledgments

I EXPRESS MY DEEP APPRECIATION TO Eric Crahan, editorial director for the humanities and social sciences at Princeton University Press, for his astute guidance and insightful advice throughout the publication process. I also thank the press for permission to use the Travel Diary and accompanying scholarly material published in Volume 14 of *The Collected Papers of Albert Einstein*, edited by Diana Kormos Buchwald, Jozsef Illy, Tilman Sauer, Osik Moses, and myself, and translated by Ann M. Hentschel and Jennifer Nollar James.

It is a pleasure to express my gratitude to my dear colleagues at the Einstein Papers Project: to Barbara Wolff for her kind assistance with my archival and library research and for her very helpful advice; to Dennis Lehmkuhl for his cheerful support and for making sure that I avoid committing any scientific howlers; to Josh Eisenthal for his friendly help with math equations; and to Emily de Araújo for her invaluable collaboration. It has been a delight working with the exceptionally competent editorial and production teams at Princeton University Press: Barbara Shi, editorial assistant, Terri O'Prey, managing editor, Chris Ferrante, designer, and Dimitri Karetnikov, illustration manager; and with Tash Siddiqui, my copyeditor, whose meticulous professionalism has been outstanding. I am also deeply indebted to Roni Grosz, Chaya Becker, and Anna Rabin of the Albert Einstein Archives for their gracious help in research and copyright issues; to historians of science Thomas F. Glick, Ildeu C. Moreira, Alfredo T. Tolmasquim, and the late Eduardo L. Ortiz for their very valuable assistance; and to Romeu Abílio, Rosemary Costhek Abílio, Ludmila Costhek Abílio, and João Samuel Rodrigues dos Santos Junior for their wise counsel. For their assistance in securing copyright permissions for the facsimiles and illustrations, I greatly thank Carolina de Paula Barbosa of the Biblioteca Nacional, Rio de Janeiro; Gabriel Feldman of Centro Marc Turkow, Asociación Mutual Israelita Argentina; Everaldo Pereira Frade and José Benito Yárritu Abellás of Museu de Astronomia e Ciências Afins; Marcelo Gleiser of Dartmouth College; Javier Rajtman of the Argentinian Amigos of the Hebrew University in Buenos Aires; Nathacha Regazzini of the Archive of Casa de Oswaldo Cruz; and Michael Simonson and Agata Sobczak of the Leo Baeck Institute. I am also deeply grateful to Dan Agulka, Bianca Rios, and Ben Perez at the interlibrary loan department at the Caltech Library for their untiring efforts in catering to my very idiosyncratic document delivery needs.

Thanks are also due to additional individuals who were instrumental in the work on the original publication of Einstein's travel diary to South America in Volume 14 of *The Collected Papers of Albert Einstein*: Andrea Labinger for her assistance with Spanish-language sources and Alexandre Cunha for his assistance with Portuguese-language sources. Work on that publication of the diary was also assisted by Mara Julseth, Herbert Karbach, Claus Spenninger, Ingeborg Wade, and the late Siegfried Grundmann.

Lastly, I am delighted to express my immeasurable gratitude to my partner Vanessa Costhek Abílio for being inspirational to me and my writing during my work on this project and for all her amazing love, encouragement, and support.

THE TRAVEL DIARIES OF ALBERT EINSTEIN

Historical Introduction

Overall impression, lacquered Indians, skeptically cynical
without any love of culture, debauched in beef tallow.

Travel diary, Buenos Aires, entry for 14 April 1925

Uruguay, happy little country, is not only charming
in nature with pleasantly warm humid climate,
but also with model social institutions.

Travel diary, Montevideo, entry for 26 April 1925

I've been roaming around this hemisphere as a traveler
in relativity for two months already. Here it's a true
paradise and a cheerful mixture of little folks.

Albert Einstein to Paul Ehrenfest, Rio de Janeiro, 5 May 1925

This Edition

This new edition presents Albert Einstein's complete travel diary from his three-month-long voyage to Argentina, Uruguay, and Brazil from March to May 1925.[1] Facsimiles of each of the pages from the original journal will be accompanied by a translation into English. Scholarly annotations will provide identifications of the individuals, organizations, and locations mentioned in the diary entries, elucidate obscure references, supply additional information on the events recorded in the diary, and offer details of his travel itinerary not mentioned by Einstein in his journal. Even though the diary constitutes a riveting historical source in which Einstein recorded his immediate impressions, by its very nature it only provides one piece of the puzzle for reconstructing the course of the trip. Therefore, to obtain a more comprehensive picture of the voyage, the annotations will draw on the contemporary local press coverages from the countries Einstein visited, additional sources from his personal papers, diplomatic reports from the time of the trip, and contemporary articles. The edition will also include a number of supplementary documents authored by Einstein which provide further context for the travel journal: letters and postcards dispatched from the voyage, and speeches he gave and statements he made at various locations.

An annotated version of this journal was originally published in its entirety in Volume 14 of *The Collected Papers of Albert Einstein* (*CPAE*) in 2015.[2] This new edition is based on the research carried out in that volume, yet presents innovative interpretations of the travel diary and its supplementary documents. For the purposes of this edition, the translation published in the English translation of Volume 14 of the *CPAE* has been slightly revised. Previously, a Portuguese translation of the travel diary had been published in 2003.[3] A Spanish translation of the Uruguay portion of the journal has also appeared.[4]

The Travel Diary

This journal is one of six travel diaries penned by Einstein. No diary is extant for Einstein's first overseas voyage to the United States in the spring of 1921. Indeed, we do not know whether he kept a travel diary on that trip.[5] The other extant diaries were written during his journey to the Far East, Palestine, and Spain, from October 1922 to March 1923, and his three trips to the United States when he visited the California Institute of Technology in Pasadena during the consecutive winter terms of 1930–1931, 1931–1932, and 1932–1933. Even though this amounts to five overseas trips, there are in fact six diaries as Einstein used two notebooks on his last voyage.[6]

The travel diary presented here was written in a notebook which consists of 72 lined pages. The diary entries appear on 43 lined pages and these are followed by 29 blank lined pages. Taking the smaller format of the notebook Einstein used for this diary into consideration, it is only a third of the size of the journal he wrote during his voyage to the Far East.

The diary provides us for the very first time with gripping insights into the most immediate levels of Einstein's experience of his journey. He wrote daily, recording initial impressions of his experiences, his reactions to the people he met, the places he visited, and the numerous official events he attended. He also jotted down pithy descriptions of the landscapes and architecture he took in, his views on the current political and social scenes, his thoughts on the local populaces and on the academic, German, and Jewish communities, and brief notes on his progress in his scientific work. And he noted his reflections on his readings during his outward voyage and occasional musings on music, culture, and contemporary world events. In addition, the diary includes drafts of poems written as dedications for portraits gifted to the three individuals of most importance to him during his stay in Buenos Aires. In contrast to the Far East diary, Einstein ceased adding entries on the last day of his South American tour and he did not continue with his journaling on the return voyage home.

The style of his diary was often very detailed, yet written in a quirky and (possibly due to lack of time) telegraphic manner. His observations of the individuals he encountered on his trip were frequently very succinct in nature—he could sum up their personalities and idiosyncrasies in just a few, often humorous or irreverent, words. Einstein kept the diary both as a record for himself and as subsequent reading matter for his wife Elsa and his younger step-daughter Margot, who remained back at home in Berlin.[7] We can be certain that he did not intend it for posterity or for publication.

The history of this journal is an intriguing one. Following Einstein's decision not to return to Germany in the wake of the Nazis' rise to power in January 1933, his son-in-law, the German–Jewish literary critic and editor Rudolf Kayser, removed Einstein's personal papers (including this travel diary) from his apartment in Berlin to the French

Embassy there and arranged for them to be transferred to France by diplomatic pouch. From there, they were shipped to Princeton where Einstein had taken up residence.[8]

In his last will and testament of 1950, Einstein appointed his loyal friend Otto Nathan and his long-term secretary Helen Dukas as trustees of his estate and Nathan as its sole executor. After Einstein's death in April 1955, Dukas also became the first archivist of his personal papers. The travel diary was one of approximately 42,000 documents transferred by the Einstein Estate to the Hebrew University of Jerusalem in 1982, in accordance with Einstein's will. This consignment of his personal papers was eventually established as the Albert Einstein Archives, initially within the Jewish National and University Library, and then, in 2008, as part of the Hebrew University's Library Authority.[9]

Background of the Trip

On 22 March 1923, Einstein arrived back in Berlin from an extensive six-month lecture tour that had brought him to the shores of the Far East, Palestine, and Spain. He returned home clearly travel-weary. Two months into his exhausting tour of Japan he had informed his sons that he was "determined not to gallivant around the world so much anymore; but am I going to be able to pull that off, too?"[10]

This travel diary is evidence of the fact that he did not manage to do so. On 4 March 1925, almost exactly two years after his return home from Japan to his apartment at Haberlandstraße 5, Einstein embarked on yet another ocean voyage that would bring him to a continent he had never visited before—South America. But what motivated him to forsake his beloved "tower room" in his attic above his spacious apartment, set out on a journey to distant lands again, cross the Atlantic Ocean for the second time in his life, and venture into the southern hemisphere?

The Genesis of the Trip

The genesis of Einstein's twelve-week trip to South America in the spring of 1925 was quite convoluted. Numerous decisive factors—from the evolution of the local scientific communities to the reception of relativity, from Einstein's political involvement to his identity as a German and a Jew, from his desire to escape from his multifold commitments in Berlin to intriguing developments in his private life—all formed the background to both the extension of the various invitations and to Einstein's own eventual acceptance of those proposals.[11]

The Early Reception of Relativity in Argentina, Uruguay, and Brazil

The initial reception of relativity in each of the countries Einstein visited was impacted more than anything else by the degree to which infrastructures existed for the exact sciences at local institutions for higher learning. In the specific context of South America, the reception process differed greatly between Argentina, on the one hand, and Uruguay and Brazil, on the other.

By 1919, Argentina already had a well-developed academic infrastructure in mathematics and physics that was based on a significant presence of scientists who were either of German origin or who had been educated in Germany.[12] Consequently, publications on relativity, mainly in French, were already being read and discussed by the local scientific community. The earliest disseminator of Einstein's theories in Argentina was the prominent writer and amateur scientist Leopoldo Lugones who gave a lecture on relativity in 1920 and published an influential book on the topic the following year.[13] This led to relativity becoming an important cultural concern among Argentinian intellectuals and several popular papers were published on the topic.[14] The local scientific community was also influenced by the Spanish intellectuals Julio Rey Pastor and Blas Cabrera who lectured on relativity in Buenos Aires in the early 1920s.[15] Local academics subsequently began to lecture and publish on relativity.[16]

In contrast to the situation in Argentina, Uruguay and Brazil did not have such infrastructure in the exact sciences and therefore the reception of relativity was delayed there.[17] At the time of Einstein's visit to Uruguay, physics was only studied at the University of the Republic's School of Engineering and at the Polytechnic Association of Uruguay.[18]

In Brazil, initial interest in Einstein's theories had been raised by the solar eclipse in 1919 that led to the verification of general relativity. Brazilian astronomers were even directly involved in the expedition to Sobral in the northeast of the country.[19] However, in spite of this early connection, in sharp contrast to the situation in Argentina, Einstein's theories only had a very limited impact on the Brazilian scientific community prior to his visit. There were two main reasons for this different state of affairs.

Firstly, Brazil did not yet have any institutions dedicated solely to research in the fields of physics or mathematics. The country's only university was that in the capital Rio de Janeiro, established in 1920, which was a cluster of schools, encompassing the Polytechnic School, the School of Medicine, and the Faculty of Law.[20] The only academics who displayed some interest in Einstein's theories were mathematicians and engineers who were self-educated in relativistic physics. The main promoters of relativity in Brazil prior to Einstein's visit were the mathematician and engineer Manuel Amoroso Costa and the mathematician Roberto Marinho de Azevedo. Costa wrote the first articles on relativity in local newspapers, held four lectures on relativity at the Polytechnic School in April and May 1922, and published the first Brazilian book on relativity based on his courses that same year.[21] Yet, apart from promoters of relativity, there were also many opponents to the new theory, particularly among intellectuals who were influenced by positivism or adhered to classical mechanics.[22]

Secondly, Brazilian scientific institutions were greatly influenced by their French counterparts and mostly established in their image.[23] This was in stark contrast to the substantial German influence on the Argentinian scientific community. French mathematician Émile Borel had lectured on relativity in Rio de Janeiro in 1922 after Costa's lecture series.[24]

Multiple Invitations from South America

As we have seen, the early reception of relativity evolved quite differently in the three countries Einstein would eventually visit. This had a major impact on the origins of the invitations extended to him by his future hosts. Yet intriguingly, in spite of the markedly different backgrounds to the invitations, there were also some striking similarities.

Undoubtedly the numerous efforts to invite Einstein to Argentina were the deciding factors in eventually bringing him to South America in the mid-1920s. It was Leopoldo Lugones, a fellow member of the International Committee on Intellectual Cooperation, and the first disseminator of relativity in Argentina, who first conceived of a lecture tour by Einstein in the country. When Lugones visited Paris in July 1921, he called on his French colleagues to send one of their physicists to lecture on relativity in Argentina. In reaction, both the German Foreign Ministry and the Prussian Ministry of Education contacted Einstein to inquire whether he himself would be willing to embark on such a lecture tour. In response, he stated that he did not foresee himself traveling to South America in the next eighteen months and recommended they invite his former collaborator, the German–Argentinian physicist Jakob Laub, in his stead.[25]

However, the assassination of the Jewish foreign minister of Germany, Walther Rathenau, in June 1922, and the ensuing personal threats on Einstein's life, led Lugones to reassess his position on the issue of whom to invite to Argentina to lecture on relativity. In the wake of the assassination, he actually proposed offering a chair to Einstein in Buenos Aires in August 1922.[26] This ambitious initiative was swiftly supported by two student unions. The following month, the Institución Cultural Argentino–Germana, the main cultural institution of the local German community, discussed the initiative to invite Einstein for a lecture tour. However, due to Einstein's reputation as a pacifist and a "traitor to the fatherland," the German members of the institution opposed the proposal.[27]

Nevertheless, after the French physicist Jorge Duclout proposed to the Science Faculty of the University of Buenos Aires (UBA) that it award Einstein a honorary doctorate and invite him to hold a lecture tour in the country, a resolution to extend such an invitation in principle was adopted in October 1922.[28] The Institución Cultural Argentino–Germana made another attempt to invite Einstein in October 1923 which he rejected due to lack of time.[29]

The invitation Einstein eventually did accept was extended to him in two stages. In November 1923, a group of prominent Jewish families invited him to visit Argentina and promised an honorarium of $4,000, yet made no mention of a lecture tour.[30] However, Einstein insisted that he could only accept an academic invitation, not one issued solely by private individuals. Therefore, the central cultural institution of Argentine Jewry, the Asociación Hebraica, informed the UBA that they would cover the cost of Einstein's honorarium and two return tickets. It seems reasonable to assume that Mauricio Nirenstein, who served as secretary of the UBA and was also closely affiliated with the Asociación Hebraica, played a crucial role in mediating between the two organizations.[31] As historian of science Eduardo L. Ortiz has pointed out, the invitation to Einstein was a very significant coup for the Asociación Hebraica and Argentine Jewry in general. It demonstrated to Argentinian intellectuals the association's ties to Europe and the Jewish community's ability to bring the most prestigious living Jewish scientist to Argentina.[32]

In December 1923, the university council of the UBA met and proposed to extend a joint invitation to Einstein to lecture in Argentina on behalf of all the country's universities. The proposed cost of the visit was projected to be almost equal to the annual salary of a top-ranking visiting professor.[33] The following month, the Asociación Hebraica informed Einstein of the official invitation by Argentina's five universities to embark on a lecture tour; the association would, for the main part, provide the funding. The invitation also stated that the University of the Republic in Montevideo would offer 1,000 pesos for possible lectures there and that an invitation to Santiago in Chile could also be arranged.[34]

Even though Einstein's identity as a scientist was the *sine qua non* condition that brought him to the banks of La Plata River, his future hosts were also very much aware of his well-established public engagement. Political factors that have been cited for the issuing of the invitation are Einstein's association with pacifism which appealed to some Argentinian intellectuals,[35] and the wish by more progressive sectors of Argentinian society to counter rising anti-Semitism which had even led to a pogrom-like massacre in early 1919.[36] Furthermore, the Asociación Hebraica seized on the invitation as a means to increase its leadership's recognition from the Argentinian intelligentsia and, more widely, the Jewish community viewed the visit as an opportunity to improve the public perception of Jews in Argentinian society.[37]

In spite of the limited impact of relativity on Brazilian science, the endeavor to invite Einstein to hold lectures in Brazil was, as in Argentina, a collaborative effort of the Jewish and scientific communities. The initiative originated with Rabbi Isaiah Raffalovich, the head of the Jewish community in Rio de Janeiro. Jacobo Saslavsky, president of the Asociación Hebraica in Buenos Aires, contacted Raffalovich to inform him that Einstein would be traveling through Rio de Janeiro and that this would be a great opportunity to invite him to lecture there. However, he also advised Raffalovich of the condition imposed by Einstein, i.e., that he would only accept invitations from official academic institutions.[38] In his reminiscences, Raffalovich revealed that part of his motivation in extending the invitation was to elevate the status of the Jewish community among their host nation: "I thought we ought to take advantage of the opportunity and demonstrate to the people of Brazil that Jews are not only peddlers but that among them one may find world famous scientists."[39] Therefore, Raffalovich urged Ignácio do Amaral, Professor at the Polytechnic School in Rio de Janeiro, to get involved.[40] An organizing committee was established to coordinate the visit which included representatives of the Polytechnic School, the Engineering Club, and the Brazilian Academy of Sciences. None of these organizations had specific ties to Einstein's theories but they were prestigious institutions.[41] As Raffalovich was the initiator of the lecture tour, he was asked by the committee to convey the invitation to Einstein. The invitation was extended on behalf of the Polytechnic and Medical Schools without any reference to the university itself.[42] A few days later, Raffalovich received a telegram from Einstein stating that he had accepted the invitation.[43]

Einstein's Acceptance of the Invitations

Einstein's own motives for accepting the various invitations from his Latin American hosts were multifold and laced with ambivalence. We have to consider the factors that led him to embark on this trip both in the wider context of his overseas travels during this period in general and the more immediate aspects of this specific tour in particular.

The broader perspectives for Einstein's decision to embark on his voyages beyond Europe in the early 1920s have been discussed by various scholars. According to German historian Siegfried Grundmann, Einstein's overseas journeys were motivated by two main factors: the dissemination of his theories and the reestablishment of international cooperation between the German and foreign scientific communities, which had been severed by World War I.[44] German historian of science Jürgen Renn has maintained that, during this period, "science became a messenger of international cooperation and Einstein its leading protagonist."[45] By late 1920, even before he had ventured beyond the European continent for the first time, in anticipation of his upcoming international trips, he was ironically calling himself a "traveler in relativity."[46]

As for the more immediate context, it was clearly the invitation from Buenos Aires that provided the tipping point for Einstein embarking on the voyage in the first place—the other solicitations clearly ensued as a consequence of the Argentinian one.

The reminiscences of prominent Argentinian astrophysicist Enrique Gaviola offer an insight into Einstein's possible state of mind after the invitation by the Asociación Hebraica and the UBA was extended. When he visited Einstein in early 1924, Gaviola presented him with a memorandum on the state of the universities in Argentina. He was dismayed to learn that Einstein was hesitant to embark on the trip as "he was pessimistic regarding the development of scientific culture in tropical countries." Gaviola explained to him that Argentina was only partially a tropical country and secured his promise "to see if he could do something useful during his visit to the country."[47] This episode provides an important insight into how Einstein perceived the state of the scientific community in Argentina prior to his tour.

Indeed, Einstein's decision to embark on the trip was definitely accompanied by a great deal of ambivalence. There was a clear discrepancy between how he expressed his views on the upcoming trip officially and privately. In March 1924, he informed the Asociación Hebraica: "This invitation delighted me so much that I most certainly feel like accepting it right away." Yet he proceeded to inform them that he could not visit in the current year due to his busy schedule.[48] However, when he shared his decision to travel to South America in

June the following year with his close friend Paul Ehrenfest, he stated that in spite of his sincere longing for "splendid isolation," he would undertake the voyage as "they [i.e., the Argentinians] are practically skinning me alive."[49] Thus, there was a definite degree of reluctance on his part. By late October 1924, he had brought the date of his departure forward, informing his son Hans Albert that he would be sailing for South America on 3 March 1925.[50]

Let us now take a look at the various incentives for Einstein to embark on this trip. First, there were the scientific factors. Brazilian historian of science Alfredo T. Tolmasquim has claimed that Einstein was motivated by his desire to meet his South American colleagues and to further "disseminate the concepts of relativity and the most current issues in physics."[51] His intention to promote relativity is also seen as a motivation for the trip by Argentinian historians of science Miguel de Asúa and Diego Hurtado de Mendoza.[52] However, the scientific incentives for heeding the call to South America are not as obvious as they may seem. We can reasonably surmise that, in light of his rather limited contact prior to his trip with Argentinian physicists and the incipient nature of the infrastructure for the exact sciences in Brazil and Uruguay, Einstein did not expect his tour to lead to new collaborative efforts on his scientific theories during his stay in South America. Therefore, there must have been other reasons for his accepting the invitations.

Indeed, as we saw in our examination of his first extant travel diary, Einstein had used the extended moratoria the ocean voyages to and from the Far East afforded him to work quite intensively on his scientific theories. He thoroughly enjoyed the isolation on board the ship and repeatedly noted in his diary how much he cherished the peace and quiet of the wide open sea.[53] In early 1923, at the beginning of his return trip to Europe, he had stated: "And how conducive to thinking and working the long sea voyage is—a paradisiacal state without correspondence, visits, meetings, and other inventions of the devil!"[54]

We can therefore safely assume that he expected to again utilize the upcoming time on board the steamships to work on and advance his scientific theories. Indeed, we have two indirect confirmations of this assumption. However, both statements also provide evidence that Einstein's enthusiasm for the upcoming trip was clearly tempered by his apprehension of the hectic social engagements planned. In December 1924, he expressed his very restrained eagerness for the upcoming trip to his sister Maja: "At the beginning of March I'll be traveling to Argentina; they've been pestering me for years, and I have now relented (out of love for the sea)."[55] And in February 1925, he shared with his friend Hermann Anschütz-Kaempfe both his positive attitude towards the planned ocean voyage and his reservations regarding the upcoming plethora of social engagements: "Long live the sea, but I'm not looking forward to the semi-cultured Indians there dressed in their tuxedos."[56] Further below we will discuss this and other less complimentary remarks on the local inhabitants he was to encounter.

Another indication that collaboration with fellow scientists was not necessarily at the forefront of Einstein's deliberations as to whether to undertake the voyage can be surmised from a consideration of an alternative invitation he rejected around this time. In November 1924, Einstein declined Robert A. Millikan's invitation to visit Caltech in 1925 due to his upcoming trip to South America. As Tolmasquim has pointed out, Einstein's adherence to his planned trip was all the more remarkable in light of the fact that the work being carried out in Pasadena was far more relevant to Einstein's own work than that being undertaken in South America.[57]

Einstein's involvement in Jewish enterprises perhaps also influenced his considerations. Tolmasquim has speculated about the possibility that Einstein's desire to get young Jewish communities in South America involved in the Hebrew University of Jerusalem played a role in his decision.[58] On the other hand, he also pointed out that Einstein forwent the inauguration of the Hebrew University, a project of utmost importance to him, to embark on the South American tour.[59] But it may even be that the upcoming opening of the university was an added incentive for Einstein to travel to Argentina, to avoid the inevitable brouhaha that would have awaited him in Jerusalem, where he had previously been feted as a national icon.[60]

Lastly, we cannot discount personal reasons that may well have been at play. Historians have speculated that, like his trip to the Far East, Einstein's interest in and fascination with new lands may have had an impact on his decision to travel. Tolmasquim, for example, has argued that Einstein was attracted to the opportunity to visit a new continent.[61] However, as we have seen above, he did not express any such enthusiasm—quite the opposite. Tolmasquim has also surmised that one reason for the delay of the trip, originally planned for the spring of 1924, was the upcoming wedding of Einstein's step-daughter Ilse, scheduled for mid-April 1924.[62]

However, it seems likely that another personal predicament had a more significant impact on both the timing of the trip and, even more so, on the fact that Einstein traveled alone. In the summer of 1923, Einstein's relationship with his then 23-year-old secretary Betty Neumann took on a more personal nature. Born in Graz, Austria, Betty was the first cousin once removed of Einstein's close friend and physician Hans Mühsam. She had started working as Einstein's secretary in June 1923. A year after the affair had begun, it was still continuing, yet had reached a critical stage. For several months, Einstein agonized over whether to break off the relationship. In June 1924, he urged her to find someone younger than himself.[63] As German science journalist Albrecht Fölsing has speculated, the desire "to put some distance between himself and his personal entanglements" may also have played a significant role in determining the timing of the trip.[64] Indeed, as we have seen, it was the very next month that Einstein informed Paul Ehrenfest that he would be embarking on the voyage in June the following year. Thus, when the timing of the journey was determined,

Einstein was in midst of repeated efforts to end his affair with Betty. The trip may therefore have provided him with a convenient way to end the relationship and also allow for Einstein and his wife Elsa to spend some time apart in the wake of what must have been a challenging period for both of them.

The affair with Betty was definitely very much on Einstein's mind at the outset of his journey. The very last letter he wrote from the European continent, as the *Cap Polonio* was about to leave Lisbon and sail for South America on 11 March 1925, was to Betty's mother Flora. In light of the fact that Einstein states that he was answering her (not extant) letter "immediately" and that it was "important to me to spare you and your daughter any disappointment," we can conclude that months after the affair had ended, Betty and her mother were still hoping for a rekindling of the relationship. In any case, Einstein clarifies

1. S.S. *Cap Polonio*, Montevideo harbor at night (Courtesy www.histarmar.com.ar).

that he cannot explain his reasons for ending the affair but that "as a conscientious and decent person I could not do otherwise than crawl away into my snail's shell."[65] Perhaps it seems fair to see the first leg of his journey—the sea voyage across the Atlantic, traveling alone in his cabin—as an expression of his strong desire to withdraw temporarily from the world and find a safe refuge.

Argentina, Uruguay, and Brazil at the Time of Einstein's Visit

In the post–World War I period, Argentina first saw years of much political and social turbulence followed by an era of greater stability.[66] The political scene was dominated by the Unión Cívica Radical party, which represented middle-class interests and advocated a liberal and patrician democracy. The country's democratic institutions were beginning a process of stabilization. In general, the political system was still one of clientelism: at both the national and regional levels, strong leaders dealt out favors to gain political support. The war and the Russian Revolution had had a significant impact on both the left and right political factions of the country and there were strong demands for social reform to combat the deteriorating socioeconomic conditions. Two violent acts of government repression against workers had occurred in the years just prior to Einstein's visit: the *Semana Trágica* or Tragic Week of 1919 (which also led to a massacre of Jewish immigrants) and Rebel Patagonia in 1922. The Catholic Church had a significant impact on the political scene and supported counterrevolutionary and antiliberal organizations. However, anarchists and communists were only a small minority and socialists and syndicalists advocated moderate reforms. In the early 1920s, Argentina was still predominantly rural with a few provincial cities. The country was dominated by its capital and main port of Buenos Aires which turned its back on its hinterland, exhibited greater affinity with Europe, and saw itself as the Paris of South America. In general, its native residents viewed themselves as superior to their fellow compatriots, newly arrived immigrants, and the rest of South America. Agriculture was the leading economic sector, with the nation's industry flailing in the postwar period. Britain and the United States were Argentina's main trading partners. Mass immigration was on the decline and the influence of immigrant societies was decreasing. There were also significant changes in popular culture. The public school system produced a highly literate population. The radio, phonograph, and cinema all helped to disseminate modern urban culture. In academia, the University Reform movement saw demands for greater openness, democratization, and modernization. At the time of Einstein's visit, Marcelo T. de Alvear was the president of Argentina. His tenure was characterized by a period of economic prosperity, nascent welfare legislation, anticorruption measures, and relative political stability despite strong internal strife within his own party.

In the first quarter of the 20th century, Uruguay underwent very significant political, social, and economic transformations.[67] The two presidencies of José Batlle y Ordóñez had brought about radical political and social reforms that modernized the country. One of the most progressive social legislation programs in the world included the eight-hour workday, trade union rights and collective bargaining, workers' compensation, protections in the workplace, the right to divorce, universal male suffrage, wide-ranging school and university reforms, and first steps towards a social security system. Batlle's Partido Colorado—literally, the Red Party—had a firm grip on power during this period. A new constitution in 1919 brought about a transition from a presidential executive to a collegiate one. However, this was mainly an instrument to solidify the one-party rule of the *colorados* or Reds. The country's other main party, the Partido Nacional, also known as the White Party or the *blancos*, formed a weak opposition. Batlle's political dominance continued even after his presidencies. At the time of Einstein's visit, José Serrato was president, who was a *colorado* but who had no close links with any of the party's major groups. Following World War I, the country experienced a renewed wave of immigration, mainly from southern Europe but also from central and eastern Europe. Most of the new arrivals settled in Montevideo. Uruguay was highly urbanized: a third of the country's population lived in the capital. The nation's economy experienced a growth of its industrial section; meat and wool were the country's main exportable products. Economic ties with Britain were on the decline but on the increase with the United States. Uruguay's economic prosperity also had a profound impact on its social indicators: the country had the lowest birth and death rates and the highest levels of literacy and newspaper readership in Latin America.

In 1925, Brazil was nearing the final years of its First Republic.[68] The country's political establishment was dominated by oligarchies of coffee plantation owners and cattle ranchers from the two most powerful states of São Paulo and Minas Gerais. It was still a predominantly agricultural society but after the First World War its industry began to expand. The Russian Revolution led to fears among the elite that the fledgling labor movement would gain wide support. There were also incipient movements for women's suffrage and for Afro–Brazilian political rights. Brazil, which had imported the largest number of African slaves during the colonial era, was the last country in the Americas to abolish slavery in 1888. However, the country did not acknowledge that it had a race problem and believed that it was an equitable multiracial society. At the same time, there was a firm belief in the country's "whitening" policy which aimed at countering the alleged danger of the encroachment of European culture by the Black and multiracial (known as "mulatto") populations by creating "a single race through the benign process of miscegenation."[69] One of the main instruments of this policy was mass immigration in the late 19th and early 20th centuries, mainly from southern Europe (and to a limited extent from Japan), which had a significant impact on the country's demography.[70] In terms of foreign policy, Brazilians sought to maintain close relations with Britain, Germany, and the United States and, at the same time, aimed to distance themselves from what they "regarded as the violent, extremely unstable and 'barbarous' Spanish American republics."[71] The term of Arthur Bernardes, who was president at the time of Einstein's visit, was a period of considerable unrest. He was deeply unpopular and "instituted an extremely harsh, repressive regime," often declaring a state of siege.[72] Middle echelons of the officer corps in the army staged repeated rebellions. Indeed, during his audience with Einstein, the president had to deal with a revolt by young members of the military.[73]

German and European Imagery of Latin America

In this section, we will summarize the findings of historical studies on the German and European imagery of Latin America. In these studies, most of the authors use the more modern term Latin America even though in practice they focus mainly on South America. We will also concentrate here on that continent.

The German and European image clusters pertaining to the South American continent and its various inhabitants have a rich history. These macro images could take on either positive or negative content, expressing admiration or contempt, sometimes both. First of all, there was the concept of the New World (which pertained also to North America). In its early positive variant, this myth viewed South America as an El Dorado, a paradisiacal locus of immeasurable natural wealth and abundance. In its early negative variant, the myth viewed the continent as a region of monstrosity, cannibalism, degraded nature, and barbarity. In a later iteration, the "new" continent was seen as an exotic, idyllic, and untamable expanse of nature marked by solitude. In a more modern guise, the image of an untamed continent morphed into a perception of South America as an endearing (or dangerous) region of frivolity and cheerfulness, a refuge for thieves, adventurers, and confidence tricksters. The continent also eventually became identified as a region in which a different economic mentality, diverse social and sexual behaviors, unfamiliar public morals, and an anthropological alterity held sway. This could also lead to an inversion of the El Dorado myth in which the allure of gold morphs into a curse.[74] Significantly, the geographic environment of the continent was perceived as having a profound impact on the intellectual capabilities and moral character of its populations.[75] These ideas derived from notions promoted by Comte de Buffon in France and Cornelis de Pauw in Germany, "that inferiority of the New World's climate with respect to Europe accounted for the retrogression of European biota—including human beings—when transported to the New World."[76]

German and European imagery of the various inhabitants of the continent also developed over time. Perceptions of the Spanish and Portuguese conquistadors and their descendants were heavily influenced by the negative image of Iberian colonial expansion, which was seen as characterized by intolerance, lust for power, genocide, and exploitation. In contrast, liberation efforts and republican aspirations of the local populations were viewed favorably by German authors. However, later political instability, militarism, and violence led to a dissipation of these positive attitudes. The indigenous populations, the Indios, were originally perceived in the 18th century as representatives of "the noble savage." In its positive variant, this image cluster led to the indigenous inhabitants being viewed as benevolent, gentle, peaceful, naïve, and sometimes beautiful. Positive communal embodiments of the "noble savage" myth were the land of the Amazons and the social state of the Incas.[77] Conversely, in the myth's negative variant, the indigenous South Americans were deemed primitive, childish, cannibalistic, and immoral. The third main population group, the African slaves and their Black descendants, were seen as victims of the Spanish and Portuguese settlers but could also be perceived as sensuous and dangerous.[78] The indigenous and Black inhabitants were often viewed as being inferior to White Europeans, who saw themselves as culturally, racially, and morally superior.[79] With the development of the English colonies in North America inhabited by White settlers, the concept of the "better American" emerged as a foil to the inhabitants of Latin America.[80] More objective, scientific perceptions of the local populations emerged in the early 19th century as a result of the explorations of Alexander von Humboldt.[81] Yet in the second half of the 19th century, adventure novels such as those of Karl May led to a reversion to a more idealized, romanticized perception of South America as a wild continent in which the rule of force, violence, cunning, and deceit dominated. In this context, the gaucho was seen as a symbol of freedom, independence, and masculinity. He embodied courage and adventure and roamed the immeasurably wide pampas. The female counterpart to the gaucho was the easygoing, instinctive, hot-blooded southern woman. In contrast to these stereotypes of the morally lax and decadent southerners, the patriotic, upstanding German was seen as remaining loyal to his principles.[82] Nevertheless, the renewed idealized perception of the continent eventually led to waves of emigration from Germany during this period.[83] In the early 20th century, some German sources began to refer to typical "Latin" characteristics. Traits such as altruism, hospitality, and the importance of family were seen as positive; yet laziness, hyperemotionality, tempestuousness, childishness, irresponsibility, and often dishonesty were seen as complementing the favorable characteristics.[84]

Stereotypical imagery relating specifically to Argentina and the Argentinians emerged in publications that described the German immigrant experience in the early 20th century. Buenos Aires, the destination of most newcomers from Germany, was seen as "uninspiring at best," even desolate, monotonous, and menacing. Even after the city experienced "flamboyant economic growth and a flourishing of the haute monde," it was characterized as displaying "tastelessness and harlotry" whose only purpose was to make money.[85] The Hispano-Argentines, in particular the *porteños*, the locally born inhabitants of Buenos Aires, were variously perceived in German sources as being resistant to change, deceitful, insincerely jovial, derivative in their attempts at high culture, cunning, frivolous, barbaric, corrupt, and lawless. In contrast, the German Michel, the quintessential German Everyman, was liable to be exploited by the locals and therefore had to discard his *Gemütlichkeit*, credulity, and easy nature. The immigrant hostility towards the local population culminated in "a single, unpleasant epithet": *Affenland*, the land of the monkeys.[86]

Analysis of the Travel Diary

In the following analysis, we will explore the deeper layers of Einstein's travel diary through a detailed examination of its text and other relevant historical sources.

We will carefully consider Einstein's perceptions of the national and ethnic groups he encountered on his South American tour and investigate these comments in the context of German and European images of that continent's local inhabitants, as revealed by historical and cultural studies. We will take a meticulous look at Einstein's preconceptions of these groups prior to his journey and the lens through which he viewed the three countries he toured. We will pay particular attention to Einstein's interactions with the Jewish, German, and scientific communities—how he viewed these specific groups and how they were situated in their respective societies. We will reflect on the general conclusions Einstein draws from the trip about the inhabitants of Europe and of both the southern and northern hemispheres of the Americas. We will consider in detail how he expressed his perceptions of all these various groups in his diary and his correspondence and whether these underwent any significant changes as a result of his new encounters.

We will also examine what this trip meant for Einstein personally. How did the trip change his own self-perceptions—as a European, a German, and a Jew? What transformations did his concepts of the Self and the Other undergo? How did Einstein's gaze—his "male gaze" and his "colonial gaze"—impact how he viewed the women and indigenous populations he met? How does the diary express his notions of national character and what explanations does he offer for that alleged phenomenon? What can studies on race and racism tell us about Einstein's views? What can we say about the nature of his travel and the personal impact the trip had on him? And how did he spend his time on board the ocean liners he sailed on, and what scientific research did he conduct during his voyages?

Furthermore, we will investigate what influence Einstein's presence had on the countries he visited. How was he perceived by the nations he toured? How did the local press react to his visit? What political and diplomatic factors were in play? How did the respective societies perceive his sojourns in the context of other prominent guests and other contemporary developments? And finally, how was relativity received in these three countries and how were the local scientific communities impacted in the aftermath of his visits?

Einstein on Argentina and the Argentinians

Perhaps the first time Einstein heard the word Argentina was from the lips of his favorite uncle Caesar Koch after the latter returned to Munich from Buenos Aires, where he had lived as a grain merchant in 1888–1889, when his nephew was merely nine or ten years old.[87] He may have regaled the young Albert with exciting stories of manly gauchos roaming the wide pampas. The boy had probably already read about these Argentinian horsemen in the tales of 19th-century adventure novelist Karl May, who played such a significant role in creating the gaucho romanticism of the Germans.[88]

Apart from this possible exotic vision of Argentina, Einstein's points of contact with the country were few prior to his departure for South America. The distant land would have represented a destination of choice (or necessity) for a couple of German emigrés with whom he had been closely affiliated. His former student and collaborator Jakob Laub became a Professor of Physical Geography at the University of La Plata in 1911. And his fellow cosigner of the "Manifesto to the Europeans," the pacifist Georg F. Nicolai, had faced increasing nationalist hostility in Germany and taken up a Professorship in Physiology at the National University of Córdoba in 1922.[89]

According to the documentation at our disposal, the only Argentinians Einstein met in person before his trip were the consul in Berlin, Alberto Candioti, to discuss a potential invitation in 1921, and the student Enrique Gaviola, who had come to convince him of the merits of a lecture tour.[90]

That would all change once he boarded the S.S. *Cap Polonio*. On the first day of his passage, Einstein settles in to enjoy the ocean voyage he has so eagerly awaited, noting "blissful peace" in his diary. The very next day, the ship picks up passengers at Boulogne-sur-Mer who seem to disturb him both auditorily and visually: "New passengers, mostly South Americans, chirping and dolled up."[91]

After two weeks on board he makes explicit references to Argentinians for the first time and they are not at all positive: "Day before yesterday, equator celebration in first class; yesterday in second class. In the former, the Argentines cut a poor figure. Rich class. Blasé, but also childish. [. . .] Argentines unspeakably stupid creatures. I'm done with them, at least,—as far as intellect and other substance are concerned—M[embers] [of the] I[dle] R[ich] C[lass]."[92]

After a very brief sojourn in Rio, the ship arrives in Montevideo and Einstein is welcomed by a small group of Argentinian academics who accompany him across the River Plate to Buenos Aires. His reaction, with a notable exception, is not favorable: "Journalists and other Jews

2. Einstein's inaugural lecture at the Colegio Nacional high school, with University of Buenos Aires secretary Mauricio Nirenstein, University of Buenos Aires rector José Arce, and Foreign Minister Ángel Gallardo, Buenos Aires, 27 March 1925 (Courtesy Archivo General de la Nación, Buenos Aires).

of various sorts, among others Nierenstein, secretary of the university. He is a good person, resigned to his fate, but the others are all more or less sordid."[93] After a delay, the ship arrives in Buenos Aires the following day. Einstein's visit is not off to a flying start: "Am half-dead from the unsavory rabble."[94]

It is apparent from his very first letter sent home to Berlin that he is pleased with his accommodation and grateful towards his hosts for the buffer they provided against the outside world: "I am lodging with a very likable family and am protected against all intrusions."[95] But at the same time, he clearly lacks enthusiasm for the planned upcoming events and does not ascribe much significance to the tour in general: "The schedule is immensely packed, but I feel strong and indifferent toward people. Because what I'm doing here is probably little more than a comedy."[96]

His first impression of Buenos Aires is rather subdued and critical: "City comfortable and boring. People delicate, doe-eyed, graceful, but clichéd. Luxury, superficiality."[97] After two days, he concludes that the city reminds him of a North American counterpart: "New York attenuated by the South. [. . .] The [Jews] want to 'celebrate' me in a mass gathering. But as I'm fed up with New York, I resolutely decline."[98] And he informs Elsa and Margot that "Buenos [Aires] is a barren city from the point of view of romanticism and intellectuality."[99]

His first meetings with university leaders are not without some positive impressions but not exactly enthusiastic: "Unassuming and friendly people without pretentiousness, but also without any sense of a mission. Sober, but they and others are genuine republicans, in some ways reminiscent of the Swiss."[100] Einstein's terse description of the first reception and introductory lecture held at a prestigious high school affiliated with the university reveals his clear disdain for the event: "Rousing speeches; I, mumbling in French amidst commotion. Philistine affair."[101] At his first scientific lecture, he expresses a more positive attitude towards the younger generation: "The young are always pleasant because they are interested in the topics."[102]

After "one week of razzmatazz" in Buenos Aires, he reports home: "This farce is actually wholly uninteresting and quite strenuous." Expressing his wish that he were already on the return journey, he resolves not to repeat the experience: "I will not embark on another such trip again, not even if it's compensated better; it's one big drudgery." And he states plainly and simply: "I don't want to be here."[103] He also generalizes his impression of the capital to encompass the whole nation: "The country here is, oddly, exactly as I had imagined: New York, mellowed somewhat by southern European races, but precisely as superficial and soulless."[104] Later he further expands this analogy between Argentina and its northern counterparts: "The newspapers are as impertinent and intrusive as in North America. Overall there are, despite the racial differences among the inhabitants, great similarities, which are explained by the intermingling of the population, the natural wealth of the country."[105] And in another comparison with the United States, he acknowledges that there is some interest in education but makes harsh judgments on which values in his mind are the most important ones in the country: "on the whole, nothing but money and power counts here, as in North America."[106]

3. Einstein in Buenos Aires, late March 1925 (Courtesy Archivo General de la Nación, Buenos Aires).

After two weeks in Argentina, Einstein states that he is "terribly weary of people. The thought of still having to gallivant about for so long weighs heavily on me."[107] By this time, he still has another two weeks left in the country and another four weeks left in South America. Yet he obviously feels he has been sufficiently long in the country to sum up his general sentiments: "Overall impression, lacquered Indians, skeptically cynical without any love of culture, debauched in beef tallow."[108]

After three weeks, the summary of his experiences for Elsa and Margot sounds more positive: "What a lot of things I've experienced! You'll read about it in my journal. All in all, it went quite well, but my head is as if stirred up inside with a ladle. If the Wassermanns hadn't protected me so well, I would surely have gone nuts; this way, only halfway so." Nevertheless, overall, he still seems to regret undertaking the tour and only the financial compensation could make up for it: "I've received a very fine payment here, so from that point of view it isn't all for naught."[109]

By the end of his time in Buenos Aires, he describes the local academics in more favorable terms: "Noon, breakfast hosted by more closely acquainted colleagues in the Tigre clubhouse." And the key figures of his stay—his hostess Berta Wassermann, his guide Mauricio Nirenstein, and the spirited writer Else Jerusalem—all receive photographs with dedication poems.[110] We can conclude that, somewhat similar to his stay in Japan, he seems to have developed a closer bond with a few individuals who played an important part in making his stay more tolerable.

Remarkably, Einstein seems to not differentiate between the members of various Spanish-speaking nations in relating to the inhabitants of Argentina. In describing an event for the inauguration of the Hebrew University, he refers to the Argentinian participants thus: "Spaniards appeared on the stage with elegant pathos. I, short speech."[111]

Even though he seems to have a general disdain for the Argentinians, there are certain ones to whom he takes a liking. Yet he clearly sees them as exceptions to the rule. On University of Buenos Aires rector José Arce he states: "Capable man. Stands out very much against his surroundings."[112] He also describes the philosopher Coriolano Alberini, dean of the humanities at the University of Buenos Aires, positively, in contrast to the "average people."[113]

Some experiences during his stay undoubtedly stand out for Einstein. He greatly enjoys his flight over the capital on 1 April: "Sublime impression, especially during ascent."[114] Yet he clearly prefers his time away from Buenos Aires. He finds "new energy" from a three-day stay at his host's country *estancia* at Llavallol, thirty kilometers south of the capital. He also appreciates the landscape of the Sierras de Córdoba mountain range: "Car ride into ancient, picturesque, sparsely vegetated granite mountains."[115] And he is impressed by Córdoba's architecture. In his mind, the city "exhibits vestiges of genuine culture with love of the soil and a sense for the sublime. Wonderful cathedral. Buildings finely proportioned (old Spanish) without daft ornamentation." Even

4. Einstein and Berta Wassermann-Bornberg, Junkers hydroplane, Buenos Aires, 1 April 1925 (Courtesy Leo Baeck Institute, New York).

though he disapproves of the city's "clerical rule," he contrasts Córdoba positively vis-à-vis Buenos Aires: "It's still better than a smug civilization without any culture."[116] Yet he finds the human interactions there almost as universally unappealing as in the capital: "Midday meal beside new governor of the province, a very refined, interesting person. Otherwise only tiresome plethora of Spaniards, journalists, and Jews."[117] He also clearly likes the architecture of La Plata: "Pretty, quiet, Italian-like city, with magnificent university buildings that are furnished in the North American style."[118]

Somewhat similar to his encounter with Japan, Einstein again becomes acquainted with a local culture through its music: "Jesinghaus showed me popular Argentine music, originating from the Incas. Naturalistic and grand. Glorious things must have perished with that nation."[119] And he writes home that "there are also more down-to-earth things, a kind of folk music that interested me very much."[120]

Einstein's opinion of the Argentinians does not change once he has returned to Europe. After docking in Bilbao, he writes to an acquaintance about his general impressions from his South American tour. However, because of the wording he uses, it seems he had the Argentinians in mind: "Over there, one finds more well-tailored clothes than fine and interesting fellows who wear them."[121] Even months after he has returned, his harsh judgment of Argentina and its inhabitants remains undiminished: "Now I have here at home a kind of small ethnographic museum, the most beautiful and loveliest things from Japan. They are a people with a deep, tender soul, in contrast to Argentina, which seems so banal and vulgar."[122] In this sentence, Einstein crossed out "South America" and wrote "Argentina" in its stead. It is certainly significant that, in his correction, Einstein changes his generalized statement about the whole continent to one focused more specifically on Argentina.

How can we interpret Einstein's mostly unforgiving assessment of the Argentinians? What is the cultural and historical context of his stereotypical remarks—both in his own writings and in German perceptions of South America in general and Argentina in particular?

In Einstein's characterizations of the Argentinians, we can detect the impact of some of the German and European image clusters pertaining to Latin America in general, and to Argentina in particular. Perhaps influenced by the inversion of the El Dorado myth, he perceives the Argentinians from the very beginning as superficial, materialistic, enslaved to luxury, and focused on external trappings such as clothing and cosmetics. He also views them as "childish," "blasé," and "stupid," thereby possibly echoing European stereotypes of the Latin American indigenous populations. More damningly, he attacks their moral integrity repeatedly, viewing them as "more or less sordid," and, in a perhaps more derogatory manner, as "unsavory riff-raff." In the past, Einstein had only used the term "sordid" (unsauber) in regard to individuals as a term to attack an opponent of his theories.[123] He had used the epithet "riff-raff" (Gesindel) to criticize anti-Semitic extremists who were threatening his life and, in his previous travel diary, to characterize local populations assailing him in the Levant.[124] As both the German terms "unsauber" and "Gesindel" have connotations of impurity, we can interpret Einstein's descriptions as xenophobic statements. In any case, we can surmise that the encounter with the Argentinians evoked strong emotions in him.

Interestingly, immediately upon his arrival, the Argentinians actually remind Einstein positively of the Swiss due to their republican spirit. However, he quickly prefers to call them both "Indians" and "Spaniards," thereby conflating two of the local populations of the continent. In neither instance does it seem that he intends these terms to be understood as favorable characterizations. In using these descriptors he highlights the Argentinians' perceived superficiality, tiresomeness, and eloquent pathos. In the past, Einstein had referred to "Indians" ironically as the custodians of "the secrets of natural living" and had used the phrase "living like the Indians" as a way to describe housing rough in a summer cottage.[125] In his previous travel diary, there are sparse hints that he regarded the Spaniards as an exotic people.[126]

In regard to Argentina, it is the locals' alleged lack of culture that evokes his harshest criticism. Even before he set out on his journey, Einstein had characterized the local population as "semi-cultured Indians [. . .] dressed in their tuxedos."[127] After two weeks in Argentina, he sums up his overall impressions in a very comparable, yet even more extreme manner: "lacquered Indians, skeptically cynical without any love of culture, debauched in beef tallow."[128] And after his return to Europe, he adds a moral dimension to the focus on external appearance: even though they may be "well-tailored," he criticizes the local population as not being "fine," a personality trait that for Einstein always alludes to personal integrity. The close similarity of these three statements enables us to conclude that he was definitely perceiving the local populace through his preconceived notions. Furthermore, in his accusations of decadence and decay, he was echoing European images of the inversion of the "noble savage" myth. These attacks of cultural hollowness and debauchery seem to be unprecedented in Einstein's writings.

He describes Buenos Aires and its residents in both positive and negative terms at first, but, ultimately, more negatively, as barren and desolate, echoing descriptions of the German immigrant experience in Argentina. In contrast, Córdoba possesses mostly positive qualities. Yet those favorable impressions do not sway him from his general attitude. Buenos Aires reminds him of New York and Argentina evokes memories of the United States. However, these associative recollections are not pleasant in nature. In both countries, only power and material wealth are of importance and they are both characterized by superficiality and soullessness. Nevertheless, for Einstein, the perceived negative characteristics of the Argentinian capital and the country as a whole are "attenuated" by its geographic location in the southern hemisphere, in an apparent adoption of positive European attitudes towards "the South." Furthermore, as we shall see again below, Einstein's encounter with South America has implications for his perception of his own continent—the negative similarities of the two American continents are greater than their differences, especially when compared to his positive view of Europe.

Einstein on Uruguay and the Uruguayans

There are no mentions of Uruguay in Einstein's writings or correspondence prior to his arrival in the country. Yet from the outset, he records his very favorable impressions: "In Uruguay I was met with genuine cordiality like seldom before in my life. There I encountered love of one's own soil without any kind of megalomania."[129] His description of the cordiality as "genuine" seems significant following a month in Argentina where he strongly criticized the perceived superficiality of its inhabitants. He is also immediately struck by the lack of exaggerated nationalism. This admiration for Uruguay's political and social conditions increases during his visit: "Uruguay, happy little country, is not only charming in nature with pleasantly warm humid climate, but also with model social institutions. [. . .] Very liberal, state completely separated from church."[130] He clearly approves of the progressive policies of the Uruguayan government and its constitution which he compares to that of Switzerland. He also evidently feels more at ease in Montevideo than in Buenos Aires: he finds the city "much more human and enjoyable" than the Argentinian capital and is keen on its picturesque harbor location and colonial architecture.[131] He finds it cozy and even the cooler weather reminds him of Europe.[132]

Einstein views the Uruguayans as "modest and natural" (again, thereby implicitly contrasting them to the Argentinians). He appreciates their not standing on ceremony but also remarks that even here

5. Einstein with welcoming committee in Montevideo, 24 April 1925 (Courtesy Leo Baeck Institute, New York).

"nothing goes without a tuxedo." He likens them favorably to the Swiss and the Dutch, relates this similarity to the comparable sizes of the countries, and advocates dividing large nations into smaller ones.[133] Einstein also regards the individuals he encounters favorably: he describes the philosopher Carlos Vaz Ferreira as a "decent, black, nervous fellow" and the engineer Carlos M. Maggiolo as a "very kind, decent person, quiet and introverted, not at all American."[134]

Einstein's positive impressions of Uruguay and its people are in stark contrast to his unforgiving views on Argentina. Crossing the River Plate he finds a very different country from the one he has just left. He clearly approves of what he perceives as its lack of ostentatiousness, its small scale, its progressive politics, and its greater authenticity. He also seems to view it as more European, both in its character and in its climate. In the context of German imagery of Latin

America, Einstein's descriptions evoke both the positive perceptions of republican efforts and the assessment of an indigenous population in alignment with its natural surroundings.

Einstein on Brazil and the Brazilians

We only have secondhand evidence for the earliest instance in which Einstein made a reference to Brazil. Philipp Frank, his successor as Professor of Physics in Prague, related in his biography the story that Einstein needed to obtain a uniform for taking the oath of allegiance before assuming his duties as an Austrian professor. The uniform resembled that "of a naval officer and consisted of a three-cornered hat trimmed with feathers, a coat and trousers ornamented with broad gold bands, a very warm overcoat of thick black cloth, and a sword." His son Hans Albert is supposed to have asked him to wear it walking through the streets of Zurich. Einstein allegedly, said: "I don't mind; at most, people will think I am a Brazilian admiral."[135]

Prior to his visit, the only other connection between Einstein and Brazil was the 1919 astronomical expedition to Sobral in the northeastern part of the country to verify the general theory of relativity. Einstein informed his mother of the success of the expeditions to West Africa and Brazil in June 1919. He mentioned their locations explicitly for the first time in a 1920 appendix to his popular exposition on relativity.[136]

Einstein does not make any direct references to Brazil or the Brazilians in his travel diary prior to his first arrival in the country. During his initial brief sojourn in Rio de Janeiro *en route* to Argentina, his recorded impressions are all favorable. He is immediately struck by the city's beautiful hilly landscape: "Sky overcast and light rain, but nevertheless a majestic impression of the bizarre giant cliffs." He also reacts favorably to the lush tropical vegetation: "Botanical garden, indeed the plant world in general, surpasses the dreams of 1001 Nights. Everything lives and thrives, so to speak, under one's very eyes." His perception of his hosts is also very positive: "My company, very cozy and pleasant." And the multiracial makeup of the local inhabitants also meets with his enthusiastic approval: "The miscellany of peoples in the streets is delightful. Portuguese–Indian–Negro, and everything in between, plant-like and instinctive, subdued by heat." The whirlwind tour provides Einstein with an overall very favorable introduction to the city and the country: "Wonderful experience. An indescribable abundance of impressions in a few hours."[137] A few days later, he reports home that, in contrast to "barren" Buenos Aires, he is "enchanted" by Rio.[138] His enthusiastic perception of the physical beauty he encounters is also conveyed in a note he composes to express his gratitude for Brazil's role in verifying general relativity: "The question which emanated from my head was answered by Brazil's sunny skies."[139]

On his arrival back in Rio six weeks later following his tours of Argentina and Uruguay, he is again greatly enthralled by the landscape: "Fantastically shaped granite-rock islands in foreground." This exoticism of the scenery is heightened by the humidity which "lends a mysterious effect."[140] His tour of the city the next day further adds to the impression that he is in an untamed location: "Next, with professors to 'Sugarloaf.' Dizzying ride above wild forest by cable car."[141] Months later, well after his return to Europe, the exotic landscape is one of his lasting favorable memories from the entire trip: "But the ocean voyage was splendid, and the Brazilian coast, with its fairytale-like forest and the conglomeration of peoples and the blazing sun."[142]

However, despite his continued positive perception of the country's physical splendor during his second sojourn in the city, Einstein now begins to also note less favorable impressions. He is struck by the alleged impact of the climate on the cognitive abilities of its inhabitants. His hosts "all give the impression of being softened up by the tropics. The European needs a greater metabolic stimulus than this eternally muggy atmosphere offers. What good is natural beauty and

28 – Março – 1925 — O Malho

ALBERT EINSTEIN,
EMULO DE NEWTON
NO RIO
DE JA-
NEIRO

As nossas gravuras mostram varios aspectos da estadia do sabio entre nós.
Einstein é o que veste roupa branca.

Passou pela nossa cidade Albert Einstein, o maior genio do nosso seculo. Na terra brasileira viu o sabio confirmada a sua theoria, em Sobral, no anno de 1919. Foi o inicio da repercussão do seu nome pelo mundo inteiro; desde 1905, porém, vinha Einstein mostrando a verdade da theoria da relatividade; uma das suas consequencias experimentaes foi o desvio da luz sob a influencia de massas de gravitação por occasião dos eclypses solares. Varias foram as tentativas para que, com precisão, fosse permittido aos sabios chegarem a um resultado positivo.

E foi no nosso ceu que isso se verificou.

6. Photomontage of Einstein's first visit to Rio de Janeiro on 21 March 1925, *O Molho*, 28 March 1925 (Courtesy Biblioteca Nacional, Rio de Janeiro).

wealth in this regard? I think that the life of a European slave laborer is still richer, above all, less dreamlike and hazy. Adaptation probably only possible at the price of alertness."[143] Einstein sees a correlation between this purported lack of incisive thinking and the Brazilians' tendency towards linguistic flourishes: "Language has greater pull [. . .] than observation."[144] In the Brazilian Academy of Sciences, this impression is deepened and he decries the prioritization of form over substance which he believes is also caused by geographical factors: "Those fellows are phenomenal speakers. When they laud someone they laud—eloquence. I believe such tomfoolery and irrelevance really does have something to do with the climate. But the people don't think so."[145]

His impressions of the individuals he encounters during his brief stay vary considerably. Among his Jewish hosts, he finds the president of the local community Isidoro Kohn "a busybody type."[146] Yet he is favorably impressed by Rabbi Isaiah Raffalovich whom he deems "very likable, intelligent, and decent."[147] Among the scientific community, he finds medical reformer Antônio da Silva Mello to be a "[c]lever, decent person" and characterizes a lunch at his house as "very cozy."[148] In contrast, he labels the head of the Medical Faculty Aloísio de Castro a "[r]eal ape."[149] He seems to be most impressed by psychiatrist Juliano Moreira whom he describes as "a mulatto and an especially exceptional person."[150] However, even though he obviously develops favorable feelings for some of his hosts, he adopts an aloof and quite patronizing stance towards the local population in general: "Here I'm some kind of a white elephant to the others, they are monkeys to me."[151] By the end of his tour he has a strong desire for the social events to be over and to not have to continue to interact with strangers: "irresistible yearning for peace and quiet from so many unfamiliar people."[152]

During his second sojourn in Rio, Einstein also relates to broader developments within Brazilian society. He is clearly impressed by

7. Einstein at the Brazilian Academy of Sciences (BAS) reception, with Juliano Moreira, vice president of the BAS, Isidoro Kohn, Henrique Morize, and Ignácio do Amaral, Rio de Janeiro, 7 May 1925, *Careta*, 16 May 1925 (Courtesy Biblioteca Nacional, Rio de Janeiro).

what he perceives as strong signs of change and progress: "It is interesting for a European to see a country which is about to establish by itself novel forms and customs."[153] He also continues to be fascinated by the multifold ethnic makeup of the local population: "Statistics on racial mixture. Blacks gradually disappearing through mixing because of mulattoes lacking power of resistance. Indians relatively less numerous."[154] He describes this rich diversity in a rather patronizing manner to Paul Ehrenfest: "Here it's a true paradise and a cheerful mixture of little folks."[155] The fact that Einstein perceives Brazil's indigenous population as primitive can be derived from his comment on the expeditions of Amazon explorer General Cândido Rondon: "His work consists of the incorporation of Indian tribes into civilized humanity without the use of either weapons or coercion of any kind."[156]

How can we assess Einstein's remarks on Brazil and its inhabitants both in the cultural and historical contexts of German and European perceptions of South America and in the context of his own previous writings?

From the very first instance in which Einstein allegedly refers to Brazil—the admiral's uniform—it is clear that he identifies the country as an exotic, outlandish location. The exaggerated comical and ornate livery is an embodiment for him of both exoticism and ostentatiousness. Upon his arrival in Rio, the "majestic," "wild," and "fairytale-like" landscape and the tropical and lush vegetation immediately confirm Einstein's association of the region with traditional European imagery of the New World. He is "enchanted" by this El Dorado-like paradise and delighted by its natural wealth and abundance. It quickly evokes for him another exotic location, the Orient.

At first, Einstein's impressions only express the positive variant of the New World myth. However, when he observes what he perceives as the deleterious effects on the region's hot and humid climate on its inhabitants, he begins to associate more negative characteristics with this natural paradise. Similar to his comments on the purported adverse effects of the local climate on the cognitive acuity of the Sinhalese in Sri Lanka in his previous travel diary,[157] Einstein again exhibits a strong belief in geographical determinism. This aligns with the historical European notion that the physical environment of South America had a decisive impact on both the intellectual capabilities and moral character of its local population. While Einstein stops short of arguing that the Brazilians are cognitively inferior to Europeans, he does imply that, due to environmental factors, they are not able to realize their full intellectual potential. He seems to argue that even a simple European laborer is capable of greater mental alertness than the most educated Brazilian. It is difficult to not view this observation as being an expression of stereotypical European superiority. Furthermore, his remarks on the Brazilians' predilection for linguistic flourishes and eloquent oratory for its own sake are somewhat analogous to his disparaging comments about the Argentinians' emphasis on external appearance. In both instances, he criticizes what he perceives as a preference for form over substance. In this, he seems to clearly favor "German inwardness" (deutsche Innerlichkeit) over more outwardly oriented forms of cultural expression that he associates with countries of the southern hemisphere.

Einstein's positive remarks on Brazil's potential for change and progress and its creation of new forms and customs evoke both the traditional German admiration for republican aspirations in South America and the European image of the New World. The specific wording he employs to articulate his gratitude for the role Brazil played in the verification of general relativity expresses his view that this "new" nation could assist in the realization of European intellectual efforts.

Einstein's perception of Brazil as an untamed, vibrant, and fecund region not only extends to its landscape and vegetation. He also regards the multiracial population as growing in a plant-like and "instinctive" manner. This comes very close to a dehumanization of the Brazilian people. This tendency is also reflected in his view of the local inhabitants as monkeys. He also refers to the residents as "little folks," a phrase which expresses both affection for and condescension towards the country's population.

In general, the ethnic diversity of the country makes a profound impression on Einstein. He is evidently delighted by the variety of the inhabitants' skin colors. Yet his remark on the alleged genetic weakness of the multiethnic inhabitants reveals both his racial worldview as well as his firm belief in the inherent characteristics of various races. Both of these ideological standpoints were also prevalent in his previous travel diary.[158]

It is perhaps noteworthy that Einstein does not refer to Brazil's colonial past. However, he does express praise for General Rondon's "civilizing" efforts among the indigenous population. This appreciation implicitly evokes the European "noble savage" myth as it infers that he perceives the Amazon inhabitants as primitive. The remarks also recall similar comments he made in his first travel diary in which he expressed approval of Britain's "enlightened colonialism" in Hong Kong.[159]

Finally, it is also important to point out that Einstein's experience of Brazil, similar to his visit to Uruguay, is limited to his touring the country's capital; in contrast, in Argentina, he actually visits two other cities besides Buenos Aires.

Einstein on the Jews in Argentina, Uruguay, and Brazil

Einstein's interactions with the Jewish communities of the three countries he visited during his South American tour were of great significance, both for Einstein himself and for the local Jewish inhabitants. What were his impressions of these communities, how did his experiences differ in the various countries, and what kinds of interactions did he have with them?

It is important to note that in two instances—Argentina and Brazil—the invitations to visit those countries were actually issued by the Jewish communities, even though they were sent on behalf of the local scientific communities. In both cases, Jewish representatives were also members of the organizing committees established to plan the local lecture tours and social events. In all three countries, Einstein was greeted formally by delegations of all the major Jewish organizations. In two of the countries—Argentina and Uruguay—Einstein lodged in the private homes of prominent members of the local Jewish communities.

At the time of Einstein's visit, the Jews of Argentina were largely an immigrant community. Following World War I, there was mass immigration, particularly from *shtetls* in Eastern Europe.[160] The predominant language and culture of the community's Ashkenazi majority was therefore Yiddish.[161] There was also a significant Sephardi minority who had mainly immigrated from Turkey, Syria, and Lebanon.[162] The two communities had little in common and developed their own distinct communal organizations.[163] The Jews mostly adopted an insular stance towards the broader society and were largely regarded as foreigners by their Gentile neighbors. There were substantial manifestations of anti-Semitism in the general population.[164] Merely six years prior to Einstein's tour, the community had experienced a traumatic event, the *Semana Trágica* or Tragic Week, in which hundreds of Jews had been killed in xenophobic rioting.[165] The younger generation was starting to form new institutions such as the Asociación Hebraica, which aimed at achieving a "cultural synthesis between Judaism and Argentineness."[166] It was this association that had invited Einstein to the country.

During his stay in Buenos Aires, numerous events were planned by the local Jewish community in honor of their illustrious guest. Einstein toured the city's Jewish quarter and visited several Jewish institutions. Following a tour of the editorial offices of a Yiddish newspaper, various synagogues, and an orphanage for girls, he noted in his diary: "The tragedy of the Jewish people: It loses its soul along with its lice. Is it similar with other peoples too? I do not believe quite as markedly."[167] This is an important statement that reflects Einstein's stance on Jewish acculturation. In contrast to his negative views on the assimilationist strivings of Western European Jewry that he perceived as "undignified mimicry," he was profoundly impressed by what he perceived as the ethnic authenticity and cultural achievements of the Jews of Eastern Europe.[168]

Einstein was not always pleased with the events planned by the local community. Shortly after his arrival in Buenos Aires, he refused to participate in a mass rally mooted by a "deputation of the Jews. The latter want to 'celebrate' me in a mass gathering. But as I'm fed up with New York, I resolutely decline."[169] Nevertheless, two weeks later he did take part in a large gathering to mark the inauguration of the Hebrew University that was organized by the local Zionist Federation and attended by 4,000 people.[170] Some of the events had amusing consequences. During his tour at the local Zionist Federation, he was shown curios: "tremendous filth was revealed under a photograph on the wall. I hope it's not to be taken as a symbol."[171] Towards the end of his stay in Buenos Aires, after a month in the country, Einstein was not hesitant to let the locals know his mind: "In the morning, very tasteless reception at the Jewish hospital. I cut those people down to size."[172] However, he was also deeply moved by some of his experiences: "Just now I'm back from a small reception by the Sephardi Jews in their temple, which was so beautiful that I had to cry. Hardly a word was spoken throughout—strange."[173]

8. Representatives of the Committee of Jewish Associations bidding farewell to Einstein, Buenos Aires, 23 April 1925 (Courtesy Archivo General de la Nación, Buenos Aires).

Einstein's support system in Buenos Aires was largely Jewish which allowed him to feel more at home in an unfamiliar environment. He lodged in the private residence of paper merchant Bruno Wassermann and both his hostess Berta Wassermann-Bornberg and his new acquaintance Else Jerusalem kept the journalists at bay.[174] It is also noteworthy that all three individuals to whom Einstein dedicated his farewell poems were Jewish.[175]

Einstein issued various statements on current Jewish issues and on Zionism during his stay in Buenos Aires.[176] He was clearly ambivalent about the attention bestowed on him by the Jewish community but also satisfied with his efforts on behalf of the Zionist movement. In late April he reported home: "Our Jews are importuning me the most

9. Einstein with Ben-Zion Mossinson and Argentine Jewish leaders. Left to right: José Lutzki, Luis Sverdlick, Natán Gesang, Isaac Nissensohn, Einstein, Mossinson, Wolf Nijensohn, José Mendelsohn, and Marcos Rosovski, Buenos Aires, April 1925 (Courtesy Centro Marc Turkow, Asociación Mutual Israelita Argentina, Buenos Aires).

with their love. I was able to be very decently effective for the Zionists. The cause is powerfully gaining ground here, too.[177]

In Uruguay, Jewish immigration, predominantly from Eastern Europe, began in the early 20th century. There was also a small Sephardi minority. Most of the Jewish immigrants were occupied in small commerce. The community developed its own organizational frameworks and maintained its own institutions, separate from the broader society. The Jews enjoyed political freedoms but did not achieve social integration—relations with their non-Jewish compatriots were only commercial and professional, yet not social. Zionist activity commenced in 1914.[178]

Among those who welcomed Einstein to Montevideo were members of a delegation from the local Jewish community and students.[179] Similarly to his stay in Buenos Aires, Einstein lodged with a Russian–Jewish family in the home of Naum Rossenblatt. The hosts only spoke Yiddish and their children spoke French.[180] Only a few official events were planned for the brief stay in Montevideo. Einstein met with representatives from the Jewish community the day after his arrival and a formal banquet was held in his honor by the community a few days later.[181]

The first Jews arrived in Brazil in the mid-16th century. Early immigration was dominated by Jews of Sephardi background, mainly in the north of the country.[182] A large influx of Jews from Eastern Europe began after World War I. It was a chain migration: newcomers arrived from the same locations and families.[183] The immigrants were confronted with a "colonially-rooted, societally ingrained antisemitism that permeated even the highest echelons of the Brazilian government."[184] Rabbi Isaiah Raffalovich, who officially invited Einstein to Brazil, bridged the gap between the Ashkenazi and Sephardi communities.[185] The early Zionist movement also formed a unifying ideology for the two communities.[186]

During his two sojourns in Rio de Janeiro, Einstein was accompanied on his various excursions by official representatives of the

Jewish community. The two different sections of the community, the Askenazim and the Sephardim, joined forces to host a large-scale reception held in Einstein's honor at the Automobile Club. Over 2,000 people attended the event. He noted in his diary: "Long speeches with much enthusiasm and inordinate adulation, but all meant sincerely." In his own speech, he emphasized the importance of Jewish solidarity and of Zionist colonization efforts in Palestine.[187] The next day, he attended receptions at both the Zionist Center and the Scholem Aleichem Library, thereby honoring the advocates of both Hebrew and Yiddish-speaking culture.[188]

Following his return to Berlin, Einstein expressed his satisfaction with his interactions with the Jewish communities he had visited. He wrote to his friend Michele Besso: "Wherever I go, I am enthusiastically welcomed by the Jews as, I am, for them, a sort of symbol of the cooperation of Jews. I thoroughly enjoy this, as I anticipate much pleasure from the unification of the Jews."[189] A month later, he informed

10. Einstein at the Jewish community reception, with Isidoro Kohn and Rabbi Isaiah Raffalovich, Rio de Janeiro, 9 May 1925, *Careta*, 16 May 1925 (Courtesy Biblioteca Nacional, Rio de Janeiro).

his sister that: "Everywhere I promoted the Zionist cause, and was received by the Jews with indescribable warmth."[190]

Einstein on the Germans in Argentina, Uruguay, and Brazil

Einstein's visits to the three Latin American countries stirred up considerable interest in their respective German communities. We have already seen that the initiative to invite Einstein to Argentina led to significant controversy prior to his visit there. How did Einstein relate to the German communities in the individual countries? What were the specific situations of their status as minorities in their host countries? How did they receive him? And what impressions of the tours did the official diplomatic representatives convey back to the German Foreign Ministry in Berlin?

At the time of Einstein's visit to the country, the German community in Argentina numbered some 30,000 people.[191] There had been a significant wave of immigration following World War I.[192] The community was not socially integrated into Argentine society, viewed the local inhabitants largely with disdain, and faced widespread hostility in return.[193] There was considerable political strife within the German community caused by the demise of the Wilhelmine monarchy and the founding of the Weimar Republic.[194] The traditionalist elite adopted a nationalistic, authoritarian, and monarchist outlook which alienated the Republicans and Socialists and "left the greater portion of the German community indifferent."[195] The local Germans established their own exclusive social circles and elite clubs which were dominated by the monarchists.[196] We have already seen that German scientists played a crucial role in the development of the local scientific community and infrastructure in Argentina which were far more advanced than their counterparts in other South American countries at the time. This formed an important backdrop for Einstein's visit.

In Buenos Aires, the controversy surrounding Einstein that became apparent in the leadup to his visit continued throughout the prominent guest's stay. On the day Einstein arrived in the capital, a Spanish translation of his pacifist article "Pan-Europe" was published in *La Prensa*. This reportedly inflamed parts of the local German community.[197] Neither the Sociedad Científica Alemana nor the Institución Cultural Argentino–Germana, which had partially funded Einstein's trip, hosted a reception for the illustrious visitor. Instead, the community focused its attention on German admiral Paul Behncke, who was visiting the country at the same time.[198]

Einstein seems to not have been perturbed by the lack of attention: "The German colony ignores me completely, which is the simplest for me; they seem to be even more nationalistic and anti-Semitic than in Germany proper. The German envoy, however, was very solicitous toward me; he is being boycotted by the Germans here because he's liberal. One generally laughs about the political foolishness of the Germans here."[199]

A week prior to his departure, the German ambassador finally hosted an official welcome. Einstein noted in his diary: "Evening reception at the German Embassy. Nothing but locals, no Germans; for the ambassador seems not to have dared to invite the latter to see me. Funny bunch, these Germans. To them I'm a stinking flower, yet they stick me back into their button hole over and over."[200] The function was attended by high-ranking Argentinian government ministers, academics, and prominent artists, officials of the German Embassy, and one sole German representative, the president of the German cultural institution Ricardo Seeber.[201] In his report to the German Foreign Ministry, the ambassador conveyed his disappointment that the local German community had boycotted all events held in Einstein's honor, "because a few of its nationalistic members disapproved of an interview by Einstein in *La Nación* as pacifist."[202]

The first German immigrants arrived in Uruguay in the early 19th century.[203] The early 1920s saw a new wave of German immigration to Uruguay, but Argentina and Brazil continued to receive far more newcomers.[204] The majority of immigrants were farmers; a small minority were merchants. Some of these businessmen achieved high status within Montevideo's social circles.[205] The German community in the capital had its own social institutions including schools, mutual aid societies, and many clubs.[206]

In contrast to Buenos Aires, the local German community in Montevideo did actually host a welcome for Einstein: "6 p.m. reception hosted by the German colony. Cozy and pleasant accompanied by coffee. Probably only the most liberal showed up."[207] However, according to the German ambassador, the Federation of German Associations had actually decided "unanimously" to greet Einstein by means of a reception committee and to hold a reception in his honor at the German Club.[208] Einstein also reported home about the more favorable welcome: "Over here the German colony is behaving more politely, after the one in B[uenos] A[ires] seriously disgraced itself by its decision to ignore me. But that just means one more burden for me."[209] In spite of the warmer welcome, an official event hosted at the German Embassy was boycotted by the local Germans as it had been in Buenos Aires: "In the evening, major reception by the German ambassador attended only by Uruguayan politicians and scholars."[210] It was during his stay in Montevideo that Paul von Hindenburg was elected German president. Einstein recorded his dismay and the reaction of his hosts: "How our paper heroes will enjoy having persuaded the plain honest German to vote for Hindenburg. It was embarrassing for the German ambassador in Montevideo; and the Uruguayans were poking fun at the Germans: The nation that had its reason beaten out of it with a stick."[211]

The impact of the German immigrant community in Brazil was more qualitative than quantitative: they imported their traditional customs and habits from their native country, lived together, applied their European experiences to their new home, and became role models. They had a profound impact on the local economy. They established a wide variety of social clubs that played an important role from the beginning of their immigration, especially for German businessmen.[212]

In Rio de Janeiro, Einstein was received with more universal acceptance. A group of German merchants invited him to a meeting with representatives of the local German community.[213] A few days later, the German ambassador hosted a "[c]ozy supper" at the Club "Germania."[214] The event was attended by the president of the German Chamber of Commerce, various businessmen, industrialists and bankers, and Isidoro Kohn. In his report to Berlin, the ambassador informed the German Foreign Ministry that the Brazilian foreign minister had been invited but had decided to send a representative instead.[215]

The German ambassadors in the three countries Einstein visited sent highly favorable reports back to Berlin. Yet they focused on the welcome he received from the host societies, not from the German communities. The ambassador in Argentina reported that the visit was "successful in every way. The guest was given the warmest of receptions and a plethora of honors from all quarters unlike that bestowed on any scholar to date." He believed that Einstein had "done more to advance interest in our culture and thereby also German prestige" than "any other scholar [. . .]. One could not have found a better man to counter the inimical propaganda of lies and to demolish the fairy tale of German barbarianism."[216] From Uruguay, the ambassador reported that Einstein's reception "by the government, the academic authorities, the populace, and the press [. . .] has hardly been given to a scholar here previously." He also expressed his pleasure that Einstein was referred to as the *sabio alemán* or German scholar in the Uruguayan press.[217] And the ambassador in Rio stated that Einstein's visit "without doubt benefited the German cause."[218]

11. Einstein at the Club "Germania" reception of the German community, with ambassador Hubert Knipping, Assis Chateubriand, and Mário de Souza, Rio de Janeiro, 8 May 1925, *Careta*, 16 May 1925 (Courtesy Biblioteca Nacional, Rio de Janeiro).

Impact of the Trip on Einstein's Perception of Europeans and Americans

How did Einstein's trip to South America reflect his stance towards the populations of the continent he was visiting and the continent whence he came? How did his tour change his attitude towards the inhabitants of these radically different regions of the world? And what impact did the trip have on his perception of the two American hemispheres?

In Einstein's writings prior to this travel diary, his ambivalent stance towards his own reference group, the Europeans, was clearly manifest. The murderous violence of World War I led to expressions of disgust at "these horrid Europeans."[219] Yet in his previous journal, the Far East travel diary, his perception of the Europeans oscillated between extremely positive and decisively negatives poles.[220] In the current journal, we have seen how he relates mostly favorably to the inhabitants of his own continent and, in many cases, unfavorably to the South Americans. Furthermore, the diary expresses his critical view of the North Americans as well. In Buenos Aires, he writes home: "The country here is, oddly, exactly as I had imagined: New York, mellowed somewhat by southern European races, but precisely as superficial and soulless."[221] In Uruguay, he refers to one of his hosts as "a very kind, decent person, quiet and introverted, not at all American."[222] He also compares the Europeans positively to the "softened up" Brazilians whose intellects are allegedly not as alert as their European counterparts.[223] Once he arrives back on European soil, he sums up his overall preference for his own continent: "I was on a lecture tour in South America and am now on the return journey [. . .]. It still is better and more interesting at home in Europe—despite the various European follies, political and otherwise."[224] And after his return to Berlin he restates his partiality towards Europe in very clear terms to his friend Michele Besso: "To find Europe enjoyable, one must visit America. To be sure, the people there are freer from prejudices, but, at the same time, mostly mindless and uninteresting, more so than here."[225] Significantly, this statement clarifies that the main reason Einstein preferred Europe was the allegedly more advanced nature of its intellectual culture.

While Einstein's private comments made in his travel diary and in his correspondence reveal his most authentic musings on Europe and America, we should also note that during this trip he also publishes an article which expresses his more public thoughts on the intellectual traditions of the two continents. In "On Ideals," published during his sojourn in Buenos Aires, he states: "the European ideal of life tends, first and foremost, to produce a 'great and unique personality,' set apart from the crowd and from the present moment. The quintessential European ideal is that of 'the hero, the fighter,' and its devotion to the world of ideals beyond the material is practically the equivalent of a 'veneration of heroes' tinged with religious overtones [. . .]. This Europeanism is already clearly evident in Hellenism. Devotion to the ideals of beauty and truth is the manifestation of an active creative spirit [. . .]. No matter how much the ideals and values of life may have changed in the history of Europe, they still retain this active, productive character." In contrast, he writes the following about the intellectual aspects of American society: "In America a high level of technical and economic acumen prevails; however, I do not believe that this excludes all spiritual life. Technical and economic action, after all, also provides room for creative gifts, since genius can overcome mechanical rules and develop them freely. Furthermore, the most rigorous organization of economic life creates possibilities for freeing spiritual creators from material concerns."[226] Even in such a public statement Einstein seems to clearly favor the European intellectual tradition over that of its counterpart in the New World.

Einstein's comments on North America in this journal reveal two more important insights. Firstly, he treats both American hemispheres, North and South America, as one continent. Even though he views the purported negative aspects of South America with less disdain than the supposed unfavorable characteristics of its northern counterpart, their similarities are still greater than their differences, particularly when viewed beside Europe. Whether its inhabitants speak English, Spanish, or Portuguese, Einstein still regards the American continent with a great deal of disparagement. Secondly, an unintended consequence of his harsh remarks about the North Americans is that they confirm, four years later, derogatory comments about the Americans that were attributed to him in a Dutch interview he gave following his trip to the US in 1921. At the time, he had tried half-heartedly to deny the remarks.[227]

Einstein's Gaze

In Einstein's travel diary to the Far East, we saw how the "two powerful objectifying gazes—those of patriarchy, the [. . .] 'male gaze,' and of colonialism—the 'imperial gaze'" manifested themselves.[228] Einstein's South American journal also provides multiple instances revealing the two gazes of this male Western traveler.

One crucial aspect of Einstein's gaze as a European touring the New World is the preconceptions that influence the way in which he views the landscapes and inhabitants he encounters. In the Far East diary, Einstein's perceptions were often filtered through European—in particular, Swiss and German—lenses. This is far less the case in this journal. The cityscape of Buenos Aires elicits memories of New York, the weather in Montevideo reminds him of Europe, and the bay of Rio de Janeiro evokes associations with the exotic Orient.

Nevertheless, as we have seen in our analysis of Einstein's perception of the inhabitants of the various countries he visits, his impressions reveal many instances in which he relates to the local populaces with European condescension. This is predominantly the case in regard to the Argentinians and the Brazilians.

He labels the Argentinians "Spaniards" and "Indians" and views them as materialistic, superficial, and enslaved to external trappings. He also deems them morally inferior. The fact that the negative characteristics of Buenos Aires are attenuated both by its southern location and by its inhabitants being mostly of southern European origin also reveals Einstein's ambivalent relationship to that part of his own continent. From his superior central European perspective, he clearly conceives of southern Europe as both alluring and inferior.

Similarly, Einstein views the Brazilians as prioritizing form over substance, yet also believes their intellects are impacted deleteriously by the tropical climate. Furthermore, he clearly associates Brazil's lush landscape and its multiethnic inhabitants with both exoticism and ostentatiousness. He has an ambivalent relationship to both those qualities. He certainly does express some affection for the Brazilians but even those instances are laced with condescension. Einstein's approval of General Rondon's enlightened "civilizing" efforts among the Brazilian indigenous population also reveals his "imperial gaze." In contrast to his general disdain for the Argentinians and his patronizing attitude towards the Brazilians, he clearly admires the Uruguayans, and, tellingly, they remind him of central and northern Europeans.

Another aspect of Einstein's gaze is how he perceives his reception by the local inhabitants. The diary includes one very revealing instance of this when he notes that he is seen by the locals as a "white elephant" and they seem like "monkeys" to him.[229] Thus, his close interactions with indigenous locals heighten Einstein's awareness of his own identity as a Westerner and as a celebrity. As E. Ann Kaplan has stated, "people's identities when they are traveling are often more self-consciously *national* than when they stay home."[230] Consequently, one important result of Einstein's interactions with locals in South America is the intensification of his identity as a European. As with his trip to the Far East, this voyage renders him even more European.

A further aspect of Einstein's gaze is its masculine quality. In the field of culture studies, the "male gaze" is a phrase used to describe the presentation of women as objects of male pleasure.[231] In this travel journal, Einstein often records his encounters with women. What insights into his male gaze do these entries provide us with?

A fundamental difference to his previous overseas trip is the fact that Einstein travels alone, without his wife Elsa, and without his stepdaughter Margot, who was going to accompany him, yet fell ill just prior to his departure.

From the very beginning of his voyage, women appear as exotic figures. In Bilbao, he notes the Spanish "women with black hair & eyes & and lace kerchiefs on their heads."[232] In Lisbon, he records a similar encounter: "Fishwife photographed with fish platter on her head, proud, impish gesture."[233]

Einstein's encounters with two individual women during his trip are of considerable significance. Early on in the sea journey, Einstein meets Else Jerusalem, an Austrian-born Jewish writer who had been living in Buenos Aires since 1911 and was returning from Europe. He is greatly struck by her and immediately gives her a telling nickname ("[u]ntamed like a panther cat").[234] It seems fair to speculate that the fact that she challenged bourgeois notions of morality and sexuality in her works appealed to Einstein. Intriguingly, he perceives her as possessing only a limited degree of femininity: "This morning, conversation with panther cat. Honest, impertinent, vain, a female only in those respects."[235] They spend a significant amount of time together, in the joint company of the captain—she reads one of her plays aloud, Einstein explains relativity to her.[236] He is clearly fond of her, yet finds her also slightly annoying: "I tease panther cat a lot, who is constantly probing me. She is amusing in her serious and impertinent way; Jewess of Russian type."[237] Upon his arrival in Buenos Aires, she plays an important role, forming a buffer vis-à-vis the multitude of welcomers and journalists: "Mrs. Jerusalem and all the stewards are standing by me."[238]

Einstein's hostess in the Argentinian capital, Berta Wassermann-Bornberg, is the other woman who plays a significant role during his stay. He describes her as a "[k]ind and cheerful housewife." In conjunction with Else Jerusalem, she forms part of Einstein's support system in Buenos Aires: "Assumes all the chicanery of a 'volunteer secretary' together with Mrs. Jerusalem."[239] Indeed, it seems that Einstein has a whole phalanx of women at his disposal to protect him from external disturbances: "Afternoon, ladies' circle ('soldiers') there, as well as German ambassador."[240] Wassermann participates in arguably Einstein's most exciting experience during his whole trip. Together they fly over the capital in a Junkers hydroplane: "Sublime impression, especially during ascent."[241]

Significantly, Jerusalem and Wassermann are two of the three recipients of personal dedication poems on the eve of Einstein's departure from Buenos Aires. By this time, his relationship with Jerusalem seems to have deteriorated: "In the afternoon, private lecture at Wassermanns' home for regiment, without panther cat. The latter *broges* [Yiddish for miffed] owing to neglect."[242] In his poem for her, he acknowledges her anger and her negative emotions towards him: "This one's for the

12. Einstein with his hostess Berta Wassermann-Bornberg and the writer Else Jerusalem, Wassermann residence, Buenos Aires, late March 1925 (Courtesy Archivo General de la Nación, Buenos Aires).

panther cat / She has slunk away in fury / Into the jungle, angry and wild / Thus this is the portrait she receives."[243]

Einstein's mention of Jerusalem in his correspondence with his wife Elsa is quite laconic: "There is a woman writer, Mrs. Jerusalem, a professor, and—a Bavarian priest. Add to that the captain, who is an uncommonly witty eccentric. Otherwise I paid little tribute to society."[244] However, he must have been aware that Elsa would read his journal upon his return to Berlin as he refers her and Margot to the diary for a detailed account of the trip.[245] In any case, in the end, Einstein seems to believe the decision to embark on the trip alone was the correct one. He writes home: "All in all, I must admit that traveling without a wife under such circumstances is simpler because there is less socializing fuss."[246]

Einstein's encounter with Else Jerusalem provides us with an interesting insight into how he deals with a creative and intellectually-minded woman unbound by bourgeois morality. Significantly, he names her "panther cat" and acknowledges her "untamed" and "wild" nature, even if, to some extent, he probably meant these epithets tongue-in-cheek. Intriguingly, even though she is of central European origin, in Einstein's perception she takes on exotic characteristics, both as a "Jewess of Russian type" and as a creature of the jungle. He also feels compelled to deny her femininity to some extent as her intellectuality does not fit squarely into his rigid concept of gender roles. However, in spite of these stereotypical images, he seems to have gone beyond a mere "male gaze" in his interactions with her, which was certainly unique during this voyage and his previous one to the Far East. Perhaps ironically, he leaves a photograph of himself behind for her to gaze at.

Einstein and the Other

What insights into Einstein's diary can cultural studies on alterity provide us with? The basis of the relationship of the traveler and the indigenous populations visited in distant lands is the Self/Other dyad.

The traveler projects a reflection of the Self onto the Other.[247] In this process, the Other is "a canvas upon which the best and worst qualities of the self can be projected and examined."[248] As Vamik D. Volkan has put it, the foreigners are "suitable targets for externalization." The traveler projects what is too painful to deal with internally onto the Other.[249] Furthermore, the Western visitor defines the unfamiliar Other "not in terms of his or her own reality, but in terms of norms promulgated by the Self." The traveler employs a filtering lens to compare "resemblances and differences between his own known, familiar world and the unknown world of the Other."[250]

How does the Self/Other dyad express itself in Einstein's journal? What qualities does he project onto the Other? How does he conceive of his own role vis-à-vis others? And how does he negotiate being in the world with unfamiliar others while he is touring distant lands?

In his Far East travel diary, we saw that Einstein often viewed himself as being assailed by the foreigners he encountered. Similarly, in the current journal, he seems to see the trip as a hardship he has to endure. Even before he embarks on his trip, Einstein conceives of his upcoming tour as something that will beset him, an ordeal others have burdened him with. He tells his sister Maja that the Argentinians have "been pestering him for years" to visit them.[251] Two days into his outgoing voyage, he has already decided that whatever happens, the outcome will not be a pleasant one: "For over there I have the choice between much pestering and agitation as a consequence of annoyance and disappointment."[252]

From the very first entry of his travel diary, Einstein is contending with his celebrity self being recognized by others: "Everyone recognizes my mug, but I'm not being bothered so far."[253] However, less than two weeks later, his attempts to create some distance between himself and others are being encroached upon. He feels compelled to hold a lecture on relativity: "Splendid isolation is crumbling." Yet at the same time, his attitude towards solitude is an ambivalent one. In spite of the greater attention, or maybe because of it, he seems to be dealing with a considerable degree of loneliness: "What a pity that Margot isn't here.

Solitude is nice, but not alone among many foreign apes."[254] He seeks out company with a few German speakers as regular conversation partners but it is clear that he does not find his existence on board among mostly non-Germans particularly appealing.

After his arrival in South America, Einstein's feeling of distance towards others which he had already tried to maintain on board the ocean liner is heightened even more. At the very outset of his tour, he employs apathy as a defense mechanism: "The schedule is immensely packed, but I feel strong and indifferent toward people."[255] Little wonder then that, eventually, he becomes increasingly drained by the company of unfamiliar and foreign others: "Otherwise only tiresome plethora of Spaniards, journalists, and Jews."[256] And the following day, he notes: "I'm terribly weary of people."[257] By the end of the tour, his desired aloofness seems to have backfired. The overwhelming feeling seems to be one of alienation: "Here I'm some kind of white elephant to the others, they are monkeys to me." To compensate, he relishes the solitary, restorative moments of the self: "In the evening alone in the hotel in my room naked I enjoy the view onto the bay with countless green, partly bare rock islands in the moonlight." [258]

During his tour, Einstein employs specific means to buffer his exposed self from others. Due to the inordinate attention afforded him, this is understandable. Various people assume the role of protector to keep the outside world at bay, at least to some extent. In Buenos Aires, this shielding role is assumed by his hostess Berta Wassermann-Bornberg, Else Jerusalem, and a "regiment" of women.[259] In Montevideo, he is guarded by two students.[260] Sometimes Einstein does not have an easy time even with the pleasant interactions. He perceives the positive attention he receives in rather negative terms, and the outright affection bestowed upon him as an engulfment of the self: "Without the Wassermanns I couldn't have made it but would have been devoured instead by sheer love."[261] He clearly regards his own role on the tour as that of a performer, invited to entertain others. Yet he does not view this role in positive terms. Shortly after his arrival, he writes home that "what I'm doing here is probably little more than a comedy."[262] By the end of the tour, he is convinced that the whole enterprise is an act of "buffoonery"[263] and he twice makes references to having "to climb back onto the trapeze."[264]

How does the strong ambivalence Einstein feels towards his fellow human beings affect the way he perceives the local inhabitants he encounters? We can surmise that he projects the following negative qualities onto the foreigners he meets during the tour: superficiality, materialism, soullessness, lack of culture, moral inferiority, and enslavement to luxury and external trappings in regard to Argentina; limited alertness and ability for incisive thinking in regard to Brazil; and eloquent pathos and prioritization of form over substance in regard to both Argentina and Brazil. In contrast, he admires the following qualities in the local inhabitants: authenticity, modesty, and a republican spirit in regard to Uruguay; and cheerfulness and multiracial diversity in regard to Brazil.

Einstein's attitudes towards his two most important ethnic reference groups, the Germans and the Jews, are also characterized by a desire for distance and reveal strong feelings of ambivalence. In the case of the Germans who boycott him in Buenos Aires, this certainly makes sense: "Funny bunch, these Germans. To them I'm a stinking flower, yet they stick me back into their button hole over and over."[265] Vis-à-vis the local Jewish communities, he exhibits both displeasure on occasion but also a considerable degree of affection, and he is even deeply moved by some of his encounters.

Einstein and National Character

As we concluded in our analysis of his travel diary to the Far East, Einstein was a firm believer in national character. These national characterizations "function as commonplaces—utterances that have obtained a ring of familiarity through frequent reiteration." Such stereotyping functions on two levels. On the more superficial level, "the discourse of national stereotyping deals primarily in psychologisms, ascribing to nationalities specific personality traits." But on a deeper level, "a nation's 'character' [. . .] is that essential, central set of temperamental attributes that distinguishes the nation as such from others and that motivates and explains the specificity of its presence and behavior in the world." During the process of assigning such essentialist qualities, "certain traits are singled out and foregrounded because they are typical" in two senses: "they are held to be representative of the type, and they are unusual and remarkable."[266]

In our study of Einstein's perceptions of the countries he visits on his South American tour and their inhabitants we have seen how he repeatedly generalizes and stereotypes the local national groups. Like in the Far East, he forms these opinions very swiftly and on the basis of his initial encounters with a very limited amount of representatives from each country. His further interactions only reinforce his preconceived notions and original impressions. This happens in spite of the fact that he actually encounters specific individuals who do not fit his stereotypical profiling of each local group but he discounts these cases as exceptions to the rule. His desire to adhere to such national characterizations is simply too strong for him to examine his own observations and conclusions critically.

The stereotyping has real consequences for Einstein and for the local inhabitants. On a personal level, its fatalistic quality seems to have a profound effect on his ability to enjoy himself on the trip. And on a more public level, it leads, at least to some extent, to his dismissal of the contributions the locals could potentially make to a wider intellectual discourse with the members of other countries.

Einstein, Race, and Racism

In his previous travel diary, we saw Einstein use racial categorization to distinguish between different races and peoples. But he actually went further than mere classification by race and commented on the biological origin of the alleged intellectual inferiority of the Japanese and the Chinese and on the environmental factors that purportedly impaired the intellect of the Indians. In the case of the Chinese, he even perceived them demographically as a threat to other races. We concluded that while he did not subscribe to a comprehensive theory of racism, he definitely made some racist and dehumanizing comments on various local inhabitants in that journal.[267]

Recent sociological studies on racism would categorize Einstein's statements as "a less coherent assembly of stereotypes, images, attributions, and explanations"[268] that he employed to justify the differences he discerned between the members of various ethnic groups. According to these studies, in such stereotyping, the "populations are represented as having a natural, unchanging origin and status, and therefore as being inherently different." Furthermore, the targeted group "must be attributed with additional (negatively evaluated) characteristics [. . .]. Those characteristics or consequences may be either biological or cultural."[269]

How does the current travel diary compare to the Far East journal in regard to racial categorization? Does Einstein continue to make comments that can be interpreted as racist? Does he argue that any of the peoples he encounters in South America are biologically or culturally inferior?

Einstein uses the term race explicitly in only one instance in his South American journal. At the National Museum in Rio de Janeiro, his hosts provide him with "[s]tatistics on racial mixture." He continues: "Blacks gradually disappearing through mixing because of mulattoes lacking power of resistance. Indians relatively less numerous."[270] He employs the word twice more in his correspondence written home during his trip, both with reference to Argentina. He describes the country as "New York, mellowed somewhat by southern European races, but precisely as superficial and soulless."[271] And he refers to its multiethnic composition: "Overall there are, despite the racial differences among the inhabitants, great similarities, which are explained by the intermingling of the population, the natural wealth of the country."[272]

There are other instances when he comments on the racial aspects of the societies he is visiting more implicitly. He approves enthusiastically of Brazil's multiracial makeup: "The miscellany of peoples in the streets is delightful. Portuguese–Indian–Negro, and everything in between, plant-like and instinctive, subdued by heat."[273] And he refers to psychiatrist Juliano Moreira as "a mulatto."[274]

It seems fair to conclude from Einstein's numerous harsh and dismissive comments on the Argentinians that he views them primarily as morally and culturally inferior. These criticisms seem to be primarily societal in nature rather than biological. However, his descriptions of them as "sordid" and as "unsavory riff-raff" (both terms that imply impurity) do seem more biological rather than merely cultural. Furthermore, he also perceives of them as "stupid" and intellectually inferior, seemingly on biological grounds. In the case of the Brazilians, he clearly ascribes their alleged incapacity for incisive thinking to the climate, i.e., environmental factors. However, his reference to the "plant-like and instinctive" nature of the multiracial inhabitants does seem like a biologically based characterization. In addition, his remark that multiracial people, or the "mulattoes," as he terms them, allegedly lack "power of resistance" towards other races reveals that he views some ethnic groups as inherently weaker than others. Intriguingly, Einstein terms the museum in Rio de Janeiro where the statistical data on "racial mixture" were imparted to him as a natural history museum, even though its official name was the National Museum of Brazil. This reveals that, conceptually, similar to many Europeans of his era, he subsumed anthropology and ethnography to natural history, thereby framing the discussion on the ethnic composition of Brazil in biological and dehumanizing terms. Moreover, as no census data regarding the racial composition of Brazil were collected between 1890 to 1940,[275] we must wonder how reliable the statistical data were that were imparted to him at the museum. It is very likely that his hosts were basing their "data" on the wishful thinking of the "whitening" policy in vogue among contemporary Brazilian elites and intellectuals, which was aimed at reducing the number of Black and multiracial people.[276] Finally, Einstein's support for the "civilizing" efforts among the indigenous population in the Amazons can also be seen as having racial undertones.

All in all, we can conclude that, similarly as in his Far East journal, Einstein continues to make comments in this travel diary which can be interpreted as racist. However, there are subtle, crucial differences. In this journal, he actually perceives of the "intermingling" of the populaces in both Argentina and Brazil as beneficial to those countries' societies. Furthermore, in this diary, he does not view any of the populations he encounters as a threat to other ethnic groups.

The Nature of Einstein's Travel

Einstein's Far East travel diary revealed that the extended sea journeys functioned as moratoria for him from his hectic life in Berlin. He used the considerable downtime these voyages afforded him with for personal introspection, intense scientific work, and relaxation. Indeed, the ocean voyages seemed to become the primary motivation for Einstein's travel: the lecture tours and social engagements became the price he had to pay to gain isolation on board.[277] We also concluded that the mode of Einstein's journeys was informed by characteristics of both the traveler and the tourist. Travel historians have differentiated between the "traveler," who is "trailblazing and self-motivated" and the "tourist," who is "reactive, following established channels and seeking prescribed experiences in predetermined ways."[278] In his independent-minded diary entries, Einstein assumed the role of a traveler. But as a VIP tourist, he was more than happy to have others arrange his itineraries for him.[279]

We have already seen that Einstein very much hoped to repeat the experiences of his Far East trip and utilize the long ocean voyages as moratoria. Does he succeed in doing so? And how does his tour of South America compare to his hectic itineraries in Japan, Palestine, and Spain? What can we learn about the nature of Einstein's travel?

Even before he sets out on his voyage to the southern hemisphere, there are crucial developments regarding Einstein's citizenships that impact the formal circumstances under which he embarks on his tour. When Einstein visited the United States and the Far East, he traveled there as a Swiss citizen. However, in February 1924, a year prior to his departure to South America, he learnt that he had become a Prussian citizen as early as 1913. Initially, Einstein was not thrilled with regaining his German citizenship, which he had renounced in his youth, in this manner. But he seems to have resigned himself to his new status.[280] Indeed, he actually applied for and received a diplomatic passport from the German Foreign Ministry in February 1925 and traveled to South America with an Argentinian visa.[281]

At the beginning of the outgoing sea voyage, Einstein enjoys the solitude. He chooses a few German speakers to converse with and also plays music with some of his fellow passengers. Similar to the sea crossing to Japan, he engages in some scientific work. However, in contrast to the previous journey, he does not seem to dedicate as much time for personal introspection—at least as far as we can tell from the diary entries. Less than two weeks into the journey, his attempts to establish a moratorium away from his everyday existence are encroached upon when he has to give a lecture to the ship's officers: "Splendid isolation is crumbling."[282] Nevertheless, he does greatly enjoy the sea passage and finds it "wonderfully relaxing."[283]

By the time he arrives in South America, Einstein declares that he feels "very equal to the exertions ahead of me and view what's ahead with serenity, although without much interest, either."[284] This lack of enthusiasm and the hectic schedule of lectures and social engagements definitely lead to his being "terribly weary of people."[285] His natural introversion would also have made the intense lecture schedule and the multitude of social engagements quite tortuous for him. In any case, by the end of the tour, he has an "irresistible yearning for peace and quiet from so many unfamiliar people."[286]

As in the Far East, Einstein demonstrates his independent-minded views through his diary entries. Only in this regard can we categorize his mode of travel as that of a traveler. As far as arrangements for the lectures and social engagements are concerned, it does seem that Einstein left these up to his hosts. Thus, the pattern established on his Far East trip, traveling in the mode of a VIP tourist, seems to be repeated during this journey as well. However, occasionally, Einstein actually rebels against the plans and refuses to collaborate, such as when he declines to participate in a mass rally planned by the Jewish community in Buenos Aires.[287] Yet in contrast to his time in Japan, he does not seem to have ventured out by himself at any stage of the visit. The fact that his wife Elsa was not accompanying him may have been a factor in this. In general, very little leisure time is scheduled into the intense tour in all three countries. Inevitably, this leads to a considerable degree of burnout that is conveyed very clearly in the final words of the journal: "Free, at last, but more dead than alive."[288] It may have been this overwhelming feeling of tedium that led to Einstein abandoning the journal at this point and not continuing with it on his homeward journey. This is in stark contrast to his previous travel diary. It seems that the moratorium of the return voyage extended to the maintenance of the journal itself.

Einstein's Scientific Research during his Trip

In accordance with his plan to utilize the ocean voyage to South America for both relaxation and scientific research, Einstein records his first work-related entry as early as the third day of the sea passage: "Idea about the foundations of Riemannian geometry."[289] He also writes home about his preoccupation with science and hints at its inevitability: "Besides that, I am busying myself leisurely with science. It turns out that I can't bear it otherwise [. . .]. When I try to leave it alone, life becomes too empty. No reading matter can substitute for it, not even scientific reading."[290] He notes further progress in three more entries during the voyage.[291] However, on the eve of his arrival in Rio, he reports home that "I didn't work a lot due to the heat, as opposed to the Japan trip."[292]

Nevertheless, Einstein does have two more opportunities to record advances in his work during his actual tour. In mid-April 1925, he informs Elsa and Margot of the connection between his most recent progress and his previous efforts: "I was on Wassermann's country estate in Llavajol for three days in complete tranquility, and had an exceedingly valuable scientific idea there that, peculiarly enough, draws on what I found during the return trip from Japan."[293] This is quite remarkable because during his hectic tours of the countries he visited in the Far East, he did not note any instances of having enough free time to ponder innovative scientific theories by himself. The only concrete research he carried out during those tours was his collaboration with his Japanese interpreter Jun Ishiwara.[294] However, two weeks after his countryside respite, on his passage from Uruguay back to Brazil, he decides that his recent progress was to no avail: "All my scientific ideas which I thought up in Argentina prove to be useless."[295]

Einstein's hint that his scientific work in Argentina was related to the research he had carried out on his homeward journey from Japan is very revealing. He had in fact worked on establishing a unified field theory of the electromagnetic and gravitational fields that was based on research by mathematician Hermann Weyl and astronomer Arthur S. Eddington. The paper he completed on the trip home was published in March 1923.[296] By mid-1923 he had abandoned this approach but the efforts on his South American voyage constituted a renewed attempt in this direction. As Einstein did not continue his travel diary on his return passage home to Germany, we do not know whether he continued work on this approach. However, it seems fair to speculate that he did so and that his pessimistic comment about the uselessness of his ideas in Argentina was premature. Two months after his arrival back in Berlin, he presented a paper to the Prussian Academy of Sciences in which he developed a new approach to a unification of the fields based on the assumption of an asymmetric metric.[297]

The Reception of Relativity during Einstein's Visits and the Impact of his Tour

In this section, we will examine the ways in which relativity was received in the South American countries Einstein toured. Extensive research on relativity's reception during Einstein's stay has been carried out by historians of science and we will summarize their findings. We will also examine what impact his presence had on the scientific and academic communities in the countries he visited.

Similarly to the newspaper reporting in other countries he visited, the Argentine press claimed that only a very limited number of "the country's scientists [. . .] was capable of following Einstein's arguments, and in a position to judge the quality of the theory."[298] Due to its complexity, it was difficult to disseminate Einstein's work in the popular press.[299] There was indeed little effort in the newspapers to explain relativity and only occasional debates on his theories, including a couple of articles written by anti-relativists such as Dante Tessieri. Unsurprisingly, the newspapers did focus on positive statements made by Einstein in which—in stark contrast to his travel diary entries—he praised the local scientific community and predicted that Argentina would have "a great economic and cultural future."[300]

Among the country's scientists, the immediate impact of his lectures was restricted to a small number of experts.[301] Nevertheless, Einstein's presence led to an increase in the prestige and status of the country's scientific community, at least for the duration of the tour. Extensive discussions on relativity did occur in the philosophical and literary journals that had limited circulation. In these, Einstein's tour was placed in the context of a series of illustrious visitors who were conferring a modern and cosmopolitan ambiance upon Buenos Aires. The visit was thus framed as part of a wider cultural phenomenon that was bringing Argentina closer to contemporary intellectual trends.[302]

One important group that had a vested interest in Einstein's visit was the Argentinian philosophical community. In their ranks there was a strong rivalry between the positivists led by José Ingenieros and the anti-positivists headed by Coriolano Alberini. The latter faction had been hoping that Einstein would condemn positivism and invited him to open the academic year at the University of Buenos Aires.[303] But he disappointed them "by dropping a text he had prepared in favour of one which avoided facing the controversies between the two camps."[304] Yet Alberini was still able to use Einstein's visit to strengthen his public arguments against positivism.[305]

In the short term, the tour does not seem to have led to major changes in regard to the specific reception of Einstein's theories: "only a few continue discussing relativity and the subject slowly died."[306] However, in the longer term, Einstein's visit did seem to have consequences for the wider scientific community. Major conceptual changes were taking place among the intellectual elites of Argentina and a gradual shift away from positivism that had been dominant for over twenty years was underway.[307] The visit "was also credited with having breathed fresh air into the country's academic circles, as well as linked with another positive development, i.e., forcing people to reflect on the nature of scientific knowledge and of physical reality."[308] There was now greater support among academics and cultural policy makers for the idea "that research in the theoretical sciences was an essential cultural element that, unfairly, had been neglected in the past."[309] This new approach led to concrete organizational changes. A department dedicated exclusively to physics was established at the Faculty of Engineering at the University of Buenos Aires and theoretical studies received particular attention.[310] Einstein's presence also had a positive impact on the younger generation of Argentinian scientists.[311]

According to press estimates, Einstein's three lectures in Montevideo were well attended with an average of 2,000 present at each.[312] However, similar to the newspaper coverage in Argentina, the Uruguayan journalists doubted that many members of the audience had any comprehension of Einstein's theories and maintained that only a limited number of experts, mainly engineers and some doctors, could understand his lectures.[313] In contrast to the virtual dearth of debate on relativity in the general Argentine press during Einstein's visit, a few Uruguayan newspapers did publish articles criticizing or refuting his theories.[314] Prominent space was given to Argentine critics of relativity, most notably Tessieri and Claro Cornelio Dassen, during Einstein's stay in Montevideo.[315] As in Argentina, Einstein's presence was seen as a boost for the anti-positivists.[316] Due to the lack of infrastructure for the exact sciences in Uruguay, there were no institutional changes in the immediate wake of Einstein's visit. On the political and social levels, the fact that Einstein "was ignored by Argentina and Uruguay high society" was attributed to the Catholic Church's influence and to anti-Semitic attitudes prevalent among the countries' elites.[317] The press also viewed Einstein as a representative of innovative ideas, "even revolutionary ideas" that would usher in a new world different from the prewar era.[318]

The Brazilian popular press delivered daily reports on Einstein and relativity throughout his stay. Lavish praise and adoration were heaped upon the prominent visitor. Yet the coverage was also riddled with misunderstandings about his theories.[319] As in Uruguay, as the exact sciences had not yet established themselves in Brazil at the time Einstein arrived in the country (with the exception of astronomy), his lectures could only be understood by a limited number of scientists. Only astronomers, mathematicians and some engineers could comprehend his theories.[320] During his stay, Einstein mainly interacted with engineers and doctors whose representatives had initiated the original invitation.[321]

Einstein's presence in Brazil stimulated the publication of many articles on his theories before and after his visit, some of which rejected relativity.[322] Prominent anti-relativists Admiral Gago Coutinho and Licínio Cardoso were present at Einstein's lectures. Like in Argentina and Uruguay, Einstein's tour highlighted profound differences between the older generation of positivists led by Cardoso and younger mathematicians headed by Manuel Amoroso Costa.[323] Cardoso published a major article attacking relativity in the popular press shortly after Einstein's departure and presented it to the Brazilian Academy of Sciences.[324] In a series of subsequent sessions at the academy, prominent opponents of Cardoso, such as Ignácio do Amaral and Roberto Marinho de Azevedo, expressed their forceful support for relativity and Cardoso became increasingly isolated. Costa, relativity's most vocal proponent in Brazil, was absent from the country during this period and did not participate in the discussions.[325]

Brazilian historians have seen the visit and the ensuing debates as part of a wider array of developments that mostly occurred in the 1920s and had a significant impact on the scientific cultural life of the country. Other such advances were the establishment of the Brazilian Academy of Sciences, the Brazilian Education Association, Rádio Sociedade, a Modern Art Week, and an interest in scientific popularization and education. These innovations also led to the emergence of a new scientific elite in Rio de Janeiro that advocated the advancement and validation of basic scientific research.[326] However, institutional changes only came about several years later when Brazil's first Philosophy and Science Faculty was founded at the University of São Paulo in 1934. It would take decades more until theoretical research groups began to focus on general relativity and cosmology.[327]

Conclusions

What general conclusions about Einstein's personality, his views and opinions, the manner in which he traveled, and the wider context of the trip can we draw from the diary he kept during his first and only foray into the southern hemisphere?

The circumstances under which Einstein embarks on this journey are quite different from his previous overseas passage to the Far East. This time there is no external threat to his life in the wake of an assassination and he expresses no longing for distant lands as had been the case with Japan. Even though this trip has been seen by historians and biographers as another one in a series of voyages Einstein undertook primarily to disseminate his theories and foster his ties with fellow physicists, there are no explicit statements by him that would confirm that his principal motivation was science-related. He expresses no curiosity or excitement about his upcoming interactions with the foreign colleagues or local inhabitants he will encounter. At best, he feels ambivalent towards the planned undertaking. He actually seems quite uninterested in (and even repulsed by) what will await him once he has arrived on far-flung shores. We must therefore conclude that his incentives for the voyage were far more personal than professional.

More than a *need* to be absent from Berlin following the slaying of a close associate, it seems Einstein was motivated by a *desire* to be away from the capital and possibly also from Elsa. In that regard, this trip has some significant similarities to his frequent absences from the apartment at Haberlandstraße 5 and perhaps belongs more to the long series of refuges from "nerve-wracking" Berlin and home life he sought in locations like Kiel in northern Germany and Leyden in the Netherlands.[328]

The broken-off affair with Betty Neumann may also have played a crucial role in the decision to embark on the trip and in its timing.

Even though Einstein displays no enthusiasm for the upcoming lecture tours, he definitely looks forward to the long sea journeys and states privately that they were his main impetus for undertaking the travel. And he seems to have thoroughly enjoyed them. Like his voyages to and from the Far East, he utilizes the ocean crossings for scientific research and relaxation, less so for personal introspection, at least explicitly. While on board, he attempts to distance himself from most of his fellow passengers, fosters closer relationships with a few German speakers, yet eventually experiences some loneliness on the large ocean liner surrounded mostly by foreigners. He seems to have missed his family members, particularly his step-daughter Margot, who was originally to accompany him.

13. Einstein in Buenos Aires, late March 1925 (Courtesy Archivo General de la Nación, Buenos Aires).

Einstein arrives in South America with definite preconceptions about the continent he is about to visit and its inhabitants. He is clearly impacted by German and European images of the region and its inhabitants prior to disembarkation and during his stay. His is a central European lens characterized by condescension and superiority and even a fair amount of arrogance. His real-life interactions with representatives of the host nations do not sway him from his prejudices, indeed they actually reinforce them. Even his positive impressions of some of the individuals he meets do not dissuade him from his stereotypical generalizations. He had assumed he would encounter exotic, ostentatious, and superficial denizens and his perceptions match his expectations. He seems to have felt vindicated by his experiences.

Einstein's aloofness, practiced carefully on board the steamer, turns into indifference just prior to arriving in South America. As a consequence of the hectic lecturing and social schedule, the apathy eventually morphs into alienation from his fellow human beings. He could not have stated it plainer than when he claims "I don't want to be here."[329]

Einstein's many protestations about not enjoying himself during the trip were possibly recorded for the sake of the two future readers of the diary—Elsa and his younger step-daughter Margot—to make them feel less abandoned while he was gallivanting around the world. Maybe there was a slight element of that in play. However, as he adheres to his harsh judgments on the inhabitants he encounters even after the trip has ended, that theory is hard to substantiate.

As with his trip to the Far East, we have seen how Einstein often forms his opinions on the locals he encounters quickly and based on first impressions which, in most cases, do not waver. Ironically, even though one of his main criticisms of the foreign inhabitants he meets is superficiality, he bases his judgments on very little evidence and without a deeper understanding of the local societies and cultures he visits.

Einstein's hasty judgments beg the question as to what underlying values informed his prejudiced opinions. Similar to his disdain for what he perceived as the alleged cognitive inferiority of the Japanese, Chinese, and Indians on his previous voyage, his condescending attitude towards the Argentinians and Brazilians seems to be primarily motivated by his intellectual elitism. On this trip too, Einstein's humanism ends when he faces purportedly impaired intellects, regardless of whether he thinks the diminishment is caused by biological or environmental factors. As with the Far East trip, he offers both inherent and cultural reasons to explain the alleged inferiority. In the case of the Argentinians, he deems them merely "unspeakably stupid." In the case of the Brazilians, he is convinced the climate is to blame. He thereby again (like in Sri Lanka) professes a profound belief in geographical determinism in spite of the opposition of his hosts. As with the claim of superficiality, this criticism may have actually been another projection on his part, as he states twice that his own brain "feels as if it's been stirred with a ladle."[330]

It is also important to note that Einstein's intellectual elitism is tied to two additional moral values of crucial significance to him: material austerity and personal integrity.

When he first encounters Argentinians on his passage to South America and lambasts their allegedly impaired cognitive abilities, he also classifies them as "members of the idle rich class" in an allusion to the social criticism of George Bernard Shaw.[331] He thereby does two things: he connects their material lavishness with their purported stupidity and he classifies the citizens of an entire nation as belonging to one supposedly unproductive social class. The backdrop to his doing so is the fact that he is traveling in the ship's first class and only perceiving a tiny minority of Argentinian society. But in doing so, he views that country's upper crust as representative of the whole country. On a deeper lever, it is also possible that Einstein felt slightly (or even very) uncomfortable traveling in such luxurious circumstances on the ocean liner and that he was projecting the very ostentatiousness he was experiencing onto his fellow Argentinian passengers. When traveling without Elsa, he usually preferred less lavish modes of transport and accommodation.[332]

Einstein continues to attack the Argentinians' subservience to luxury even after his arrival in the country and sees this as one of the features of their national character. There seems to have been a form of inverted snobbery in play here and it has far-reaching consequences for his perception of all the inhabitants of Argentina.

We have also seen that Einstein views the Argentinians as lacking in moral integrity. He attacks them as "more or less sordid" and "unsavory." He thereby ties personal decency to images of uncleanliness and seems to adopt certain aspects of an adherence to bourgeois respectability. This emphasis on hygiene also aligns with his biological worldview that we have already encountered in his previous travel diary.[333]

Einstein's criticism of what he perceives as the Argentinian and Brazilian tendency to prioritize form over substance (whether in attire or rhetoric) seems to have been based on a patent intolerance for expressions of culture that were dissimilar to his own. It seems as though in his mind, only certain, restricted cultural modes are legitimate and authentic. When he detects a lack of his own inward-looking standards in other groups' forms of self-expression, he deems those collectives to be morally corrupt or inauthentic or both. Here again, there is a distinct superficiality in play on Einstein's part. When the cultural forms are predominantly of an outward nature, he focuses on the external trappings and refrains from taking a deeper look beneath the surface.

In contrast to his harsh comments on the Argentinians and his patronizing remarks on the Brazilians, Einstein expresses only admiration for the Uruguayans, the small scale of their country, and their progressive political and social institutions. Unsurprisingly, they remind him the most of his fellow Europeans.

As we have noted, this was Einstein's second voyage to the American double continent. And its southern hemisphere often reminds him of its northern counterpart. From the aloof and distant perspective of an elitist central European, this is perhaps understandable. As a result of his tour, he comes to view its two parts as one whole entity, a foil for his own continent which he perceives in a more favorable light as a result of his voyage. His castigating remarks about North America also confirm the controversial criticisms attributed to him four years earlier in the aftermath of his trip to the United States.

As for the nature of Einstein's travel: similar to his previous overseas journey, Einstein again adopts a rather passive role as a VIP tourist. Tellingly, he is even traveling on a diplomatic passport. He is certainly happy to surrender his autonomy for greater convenience and leave logistical arrangements to his scientific and Jewish hosts, even though, in the end, this leads to his frantic schedule and his eventual burnout. In contrast to his Japan tour, where he found accommodation in hotels and inns, Einstein stays in private lodgings for most of his South American stay. It is no coincidence that he sojourns with Jewish families. No doubt this contributes to his feeling more at home in strange surroundings. His hosts do their best to insulate him from external intrusions and he greatly appreciates their efforts. He has a well-developed support system in the various locations he visits, most notably in Buenos Aires where he has a whole array of women protecting him. In a way, this replicates his home situation in Berlin where Elsa would keep unwanted intruders at bay. These mechanisms must have suited Einstein's introverted and aloof personality well, even if they had limited effect during his tour.

During his previous overseas journey, Einstein had been hosted by Jewish communities in Singapore, Hong Kong, Shanghai, and, of course, Palestine, but not in Japan, the principal destination of his trip. Conversely, he did meet with representatives of the German community in Tokyo. In striking contrast, on his tour in South America he has to contend with both Jewish and German communities. In the case of the Jewish residents, a considerable amount of his time is taken up by events organized by their representatives. Einstein welcomes the attention in general and, in some instances, is even moved by his encounters. Overall, he feels overwhelmed by the outpouring of affection bestowed upon him, yet pleased with his efforts on behalf of the Zionist cause and encouraged by signs of Jewish solidarity. In the case of the German communities, he is struck by the differing reactions towards him—rejection in Argentina, an ambivalent reception in Uruguay, and acceptance in Brazil. In light of the restrained enthusiasm with which he is welcomed by the local German immigrants (as opposed to the German diplomats), it is no wonder that he views them with detachment and irony.

We have seen how Einstein's gaze plays itself out during this trip. Similar to his perceptions on the Japan trip, his "imperial gaze" leads to condescending attitudes towards the local inhabitants he encounters. However, unlike in his previous diary, there are no expressions of extreme misogyny in this journal. He develops an intriguing relationship with an intellectually-minded woman he meets on board the ocean liner and one must wonder how Elsa would have reacted back at home to descriptions of her husband's flirtatiousness with her near-namesake upon reading the diary. One cannot escape the impression that Einstein may have intentionally wanted to provoke her.

Intriguingly, Einstein perceives of his own role on this trip in quite a different light than during his journey to the Far East. On that voyage, possibly in light of his temporary escape from Germany in the wake of Walther Rathenau's assassination, he viewed himself mainly as a protagonist who had to prove himself against multiple challenges from various quarters. He seemed to be under frequent attack. Here he assumes less the role of a valiant hero and more the part of a circus performer. Even before his arrival, he is entertaining the ship's officers with a lecture on relativity. He comes to see the tour and the social engagements as a "farce" and "little more than a comedy." In his own eyes, he is the trapeze artist who has to put on a show again and again for his adoring audiences. There is a considerable degree of self-mockery in these remarks and we must admire his sincerity in recording them.

As with Einstein's disturbing xenophobic and, at times, racist comments expressed in his previous travel diary, we are again confronted in this journal with his unpleasant and offensive remarks on the various local inhabitants. On multiple occasions, he comes close to dehumanizing the populaces he encounters, comparing them to plants and animals. His evocation of a racist stereotype in regard to his own ethnic group (his reference to the alleged lice of unassimilated Jews) does not mitigate the impact of his prejudiced observations directed against others. It simply reinforces the impression that his racial worldview extended to the members of all nationalities. We have noted, that, crucially, there are subtle differences in his view on races in this journal in comparison to the Far East diary: he expresses himself positively on the beneficial impact of the multiracial compositions of the societies he visits and does not include any allusions to feeling threatened by the members of another race. At times, his racist comments are casual and even cozy in tone and laced with humor. Nevertheless, we still have to contend with the harsh, disturbing, and sweeping character of his remarks. Yet without condoning the distasteful nature of his observations, perhaps we can be grateful to Einstein for his authenticity and honesty, for his unguarded remarks intended only for a very limited readership. They offer us the opportunity to deal with the fact that even the most revered human beings have a darker, more primal side that we cannot and should not ignore or discount.

Furthermore, the obvious limits of Einstein's humanism, expressed in his candid comments in his previous travel diary and now in this journal too, challenge us to confront the more instinctual sides of our own personalities and to examine our own deep-seated prejudices and biases. In both journals we are confronted with the stark contrast between Einstein's public liberalism and his private illiberalism. Looking forward from this travel diary, we must also wonder whether Einstein's intimate thoughts will change with regard to the representatives of other nations in his later years, when he begins to publicly oppose racism, e.g., in his statements against the discrimination of African–Americans in the United States. This will constitute a fascinating question when we explore the diaries from his three journeys to the US.

Like on his previous trip, Einstein reaches a point when he declares that he will not embark on such a journey again. In the former instance, he actually expressed some doubt as to whether he would adhere to his own intention. However, in this journal, the pronouncement sounds more definitive and final. Indeed, it will be another five years before he ventures overseas again, for the first of his three American trips in the early 1930s. His prolonged illness in 1928 presumably contributed to his refraining from such strenuous travel in the interim.[334]

Towards the end of his exhaustive and exhausting tour of Japan, Einstein dramatically stated that "I was dead, and my corpse rode back to Moji."[335] The final words of his South American diary definitely echo that colorful description: "Free, at last, but more dead than alive."[336] Einstein was clearly pushed to the limit on both occasions and one cannot help but wonder what toll these journeys had on him and his health.

In the end, we are therefore left with an inescapable conundrum. Why does Einstein embark on a trip that will provide him with two very dissimilar and contradictory experiences: sea journeys that constitute prolonged moratoria from his multifold obligations in the German capital and from a marriage that is clearly in troubled waters as a result of a recent affair on the one hand, and hectic lecture tours and social engagements that will stretch his introverted self to its utmost limit on the other? There seems to be no clear answer to this perplexing question, at least if we focus on Einstein as a purely rational and cerebral human being. But of course such a perception would be incomplete at best and would deny him a crucial part of his humanity. Like all of us, Einstein was sometimes driven by irrational forces and conflicting factors. Even though he desperately yearned for a safe refuge, he knew his encounters on foreign shores would not provide him with one. Perhaps he had to prove this to himself. In any case, similar to us all, Einstein had the right to be wrong and to make potentially self-sabotaging decisions. This only renders him more human.

Südamerika 1925

1

TAGEBUCH

Diary trip South-America,1925

Item 2 – [illegible] + [illegible].

No. 86 lin.

Travel Diary
Argentina, Uruguay, Brazil[1]

5 MARCH–11 MAY 1925

[Hamburg] 25 2

5. III.

Gestern Frau, Katzenstein, mit Schwester Bärwalds an der Bahn. Reise bei Sonnenschein. Ankunft trübe. Fr. Robinow mit Schwiegersohn an der Bahn. Nachmittag mit Frau Robinow auf Kindervioline Mozart. Abends hamburgerisches Familienessen. Melchior auch erschienen, nüchtern, witzig. Hotel zu Fuss. Gesellschaft nüchtern, anständig.

Heute um 9 Uhr aufs Schiff mit Schwiegersohn R. Feiner, intelligenter Mensch. Abfahrt 9 1/2 bei Sonnenschein an Schiffen Speichern vorbei. Alles Abschied. Alle kennen meine Visage, aber bisher ungestört. 11 Uhr Himmel trübe. Ufer weichen zurück. Glückselige Ruhe

Gestern Abend schwarze Krawatte ins Hotel erhalten, von Bärwalds telephonisch besorgt. Humorvolle, ungeheuer bewegliche Kerle.

5 March.

Yesterday wife, Katzenstein with sister, the Bärwalds, at the railway station.[2] The sun was shining during the journey; arrival overcast.[3] Mrs. Robinow with son-in-law at the station.[4] Afternoon with Mrs. Robinow, [played] Mozart on child's violin. In the evening Hamburg-style family meal. Melchior also appeared,[5] clear-headed, witty. Hotel by foot. Society sensible, decent.

Today at 9 a.m. boarded ship[6] with son-in-law R[obinow].[7] Fine, intelligent person. Departure 9:30 in the sunshine, past ships, warehouses. It's quite a farewell. Everyone recognizes my mug, but I'm not being bothered so far.

11 a.m. sky overcast. Shores retreat. Blissful peace

Yesterday evening received a black necktie sent to the hotel, ordered by phone by the Bärwalds. Humorous, terrifically agile fellows.

3

Kabine luxuriös. Maschine nicht zu spüren. Schiff riesig (23000). Hätte Margot gefallen.

6. III.

Boulogne Hafen. Immer trübe, aber die Luft schon mehr schmeichelnd. Neue Passagiere, meist Südamerikaner quatschend und herausgeputzt. Hafendampfer macht elegantes Abschiedsmanöver. Stadt grüsst aus Nebel herüber. Habe die Bekanntschaft von Prof. Jesinghaus (Psychologen) gemacht, der zu den Mahlzeiten bei mir sitzt. Still und fein, weiss auch ziemlich viel. In den Zwischen-Zeiten bin ich allein und recht behaglich. Reise dauert gottlob noch recht lang, aber mir graust schon vor der Zukunft.

Cabin luxurious. Can't feel the engine. Ship huge (23,000). Margot would have liked it.[8]

6 March.

Boulogne harbor. Continues to be overcast but the air is already more balmy. New passengers, mostly South Americans, chirping and dolled up. Harbor steamship makes elegant farewell maneuver. City extends greetings through the mist. Made the acquaintance of Prof. Jesinghaus (psychologist),[9] who sits near me at meal-times. Quiet and gentlemanly, knows quite a lot, too. In between meals I am alone and quite comfortable. Voyage will last quite a while yet, thank God; but I already dread the arrival.

7. III 4

Gespräch mit Jesinghaus über Kausalität.
Idee über Begründung der Riemann'schen
Geometrie. Lektüre feiner Geschichten
von Chaucer.

8.

Das erstemal strahlende Sonne.
Gegen 11 Uhr näherten wir uns Bilbao
Hel blaugrün Ufer mit Hügeln erst
silbrig dann in strahlender Sonne.
Viel Neugierige kamen aufs Schiff
Spanier neugierig, nicht ermüdet oder
blasiert, kindlich, selbstbewusst,
die Frauen mit schwarzen Haaren u. Augen
u. Spitzentüchlein auf dem Kopf
Ich genoss die Sonne dankbar,
ganz oben. Frau Jerusalem kennen
gelernt. Urwüchsig wie Panterkatze
9 Uhr abends im Zimmer für Kinder und
Frauen Klavier gespielt. Von zwei Schwäbinnen

7 March.

Conversation with Jesinghaus about causality. Idea about the foundations of Riemannian geometry.[10] Perusal of decent stories by Chaucer.[11]

8.

First time brilliant sunshine. Around 11 a.m. we approach Bilbao. Sea blue-green, shoreline with hills first silvery, then in brilliant sunshine. Many inquisitive people came on board ship. Spanish inquisitive, not weary or blasé, childish, self-confident, the women with black hair & eyes & and lace kerchiefs on their heads. I gratefully enjoyed the sun, on the very top deck. Met Mrs. Jerusalem.[12] Untamed like a panther cat. 9 p.m. played piano in room for children and women. Chased away by two Swabian women

5

vertrieben, die wohl nur zuhören wollten. Ich muss drollige Figur gemacht haben bei der Flucht.

9.

Fast den ganzen Tag in der Sonne. Morgens Landung in Coruña, nachmittags in Vigo. von hügeligen Inseln ~~[illegible]~~ abgeschlossene Bucht mit malerischem, terrassenartig gebautem Nest. (Nachmittags Gespräch über Logik). Farbenpracht und Sonnenuntergang in Vigo unvergleichlich. Die Sonne des Südens berauscht

Heute Morgen Gespräch mit Panterkatze. Ehrlich, impertinent, eitel, unklaren Weib.

11.

Gestern mit Jesinghaus und Sievers Streifzug durch Lissabon. Macht verlumpten aber sympathischen Eindruck. Gemütlich, gutmütig, ohne Eile oder auch nur Ziel

who probably just wanted to listen. I must have cut a droll figure making my escape.

9.

Spent almost all the day in the sun. Morning landing in Corunna, afternoon in Vigo, bay closed off by hilly islands with picturesque, terraced hamlet.[13] (Afternoon conversation about logic.) Blaze of colors and sunset in Vigo incomparable. The sun in the south is intoxicating.

This morning, conversation with panther cat.[14] Honest, impertinent, vain, a female only in those respects.

11.

Yesterday with Jesinghaus and Sievers,[15] ramble through Lisbon. The impression it makes is rundown but likable. Life seems to unfold cozily, good-naturedly, without haste or even a goal

6

und Bewusstsein scheint sich das
Leben abzuspielen. Allenthalben wird
man alter Kultur bewusst. Grazie.
Fischverkäuferin photographiert
mit Fischplatte auf dem Kopf, stolzer
schelmischer Geste. Besuch auf Kastell
mit imponierender Aussicht auf
Stadt und Hafen. Dann im Auto
zu nahem Kloster am Meere. Ganz
spielerische Spätgotik. Weitausl. elliptisches Gewölbe
mit palmartigen Ansätzen und
oberer Ausstrahlung wie Palmkrone.
Wundervoller Kreuzgang in zwei Stock-
werken. Drolliger Brunnenlöwe. Dann
aufs Schiff zurück. Dies verlumpte
Land flösst mir eine tiefe Sehnsucht
ein.

Heute schon gehörig warm bei
bedecktem Himmel.

or awareness. Everywhere, one becomes conscious of ancient culture. Grace. Fishwife photographed with fish platter on her head, proud, impish gesture. Visit up to castle with impressive view onto city and harbor.[16] Then in car to nearby monastery by the sea.[17] Very playful late Gothic. Expansive elliptical vault with palm-like appearance and upper radial spread like the crown of a palm tree. Wonderful cloisters in two stories. Droll fountain lion. Then back on board ship. This run-down country instills in me a kind of yearning.

Today already well and truly hot with overcast skies.

7

12.

Schon morgens (bei ziemlich heiterem Himmel) so warm, dass man nicht fühlt, ob Kabinenfenster offen ist. Lese Meyerson. Geistreich, aber insofern ungerecht, als die Eskapaden von Weyl und Eddington zum Wesen der Relat. Theorie gerechnet werden. So kommt er zum Vergleich mit Hegelei. Habe gestern bewiesen, dass bei feststehenden Lichtkegeln und Elektronenbahnen (von allen $\frac{\varepsilon}{\mu}$) bei festliegenden Koordinaten keinerlei Veränderung des Feldes möglich ist, falls man die Gleichungen für Elektronenbahnen aus der Extremaleigenschaft von

$$\int \left(g_{\mu\nu} \frac{dx_\mu}{ds} \frac{dx_\nu}{ds} + \varphi_\mu \frac{dx_\nu}{ds} \right) ds$$

ableitet.

Längs Teneriffa gefahren. Pik im Sonnenglanz. Wunderbare Beleuchtung

12.

In the morning already so hot (with quite clear skies) that one doesn't feel whether cabin window is open. Am reading Meyerson.[18] Full of wit but unfair insofar as the escapades of Weyl and Eddington are considered essential parts of the theory of relativity. That's how he arrives at the comparison with Hegelianism. I proved yesterday that given fixed light cones and electron orbits (of all $\frac{\varepsilon}{\mu}$) with defined co-ordinates, absolutely no change in the field is possible, if one derives the equations for the electron orbits from the extremal property of $\int \left(g_{\mu\nu} \frac{dx_\mu}{ds} \frac{dx_\nu}{ds} + \varphi_\mu \frac{dx_\nu}{ds} \right) ds.$[19]

Sailed past Tenerife. Peak in glowing sunlight. Wonderful illumination

8

der grünen steilen Höhen.

13.

Mittagessen mit Panterkatze und Kapitän. Viel Humor und auch Ernst. Jüdin und Ostdeutscher, beides feste Exemplare. Hat 3 Stunden gedauert. Habe Kriegers Buch zu lesen begonnen. Merkwürdiges Manövrieren mit abstrakten Begriffen, die für östliche Menschen wohl auch einen Gefühlsgehalt haben. Weniger als 20° nördliche Breite bei ziemlich frischer Temperatur und starker Brise.

14.

Vormittag Musizieren mit einer jungen Witwe und Jüngling, es war sehr mässig. Nachmittags liest Frau Jerusalem bei dem sehr netten Kapitän ihr Drama vor. Konflikt zwischen jüdischem Milieu

of the steep green mountains.

13.

Lunch with panther cat and captain. Much humor and earnestness, too. Jewess and east German, both solid specimens. It lasted three hours. Began to read Koigen's book.[20] Strange maneuvering with abstract concepts that for people of the East surely also have emotional content. Less than 20° northern latitude at quite cool temperature and strong breeze.

14.

Before noon played music with a young widow and young man; it was very mediocre. In the afternoon, Mrs. Jerusalem reads her play aloud in the presence of the very nice captain. Conflict between Jewish milieu

9

und einen größeren Geschlechts-
und Wirkungskreis suchenden
Sohn, eine Jesusgestalt. Allzu
abstrakt aber doch packend.
Erklärte der Frau Grundgedanken
der Relativität.

Morgens wurde mir Geburts-
tagsbriefchen gebracht. Ich war
doch gerührt. Abends Gespräch
über Wesen der Religion mit Jesing-
bei wundervollem Sternhimmel
haus. ~~Sonnenuntergang~~ bei
wunderbarer Beleuchtung der
(Kap Verd. I.)
Insel Fogo. Die Spitze der jäh
aufragenden Gebirgsinsel glühte,
das übrige in mattem Blau.

15.

Erster Tropentag. Matt aber doch
angenehm bei bedecktem Himmel.

and a son seeking a broader horizon and sphere of activity, a Jesus figure.[21] Overly abstract but gripping nonetheless. Explained to the woman the basic principle of relativity.

In the morning, a small birthday missive was brought to me.[22] I was touched nevertheless. In the evening, conversation about the essence of religion with Jesinghaus under wonderful starry skies. Sunset with wondrous illumination of the island Fogo (Cape Verde). The tip of that precipitously soaring mountain island glowed, the rest in matte blue.

15.

First tropical day. Languid yet still pleasant with overcast sky.

10

12.

Heute Aequatorfest mit Neptun
und Taufe. Viel harmloser Jux.
Habe verschiedentlich musiziert
mit der fidelen Witwe Ohnesorg
und einer jungen bemalten
Chilenerin. Denn man muss am
19. bei Fest mitmachen. Muss
auch den Offizieren Rel. Vortrag
halten. Splendid Isolation
bröckelt ab. Aber die Reise
dauert auch nicht mehr
ewig. Hitze 22 Ream. im Schatten
Nachts gehörig schwitzen
Habe Überzeugung gewonnen,
dass $(R_{ik} - \frac{1}{4} g_{ik} R) = T_{ik}$ nicht
das Richtige ist. Überzeugung
von Unmögl. von Feldtheorie im
bisherigen Sinne verstärkt sich.
Schade dass Margot nicht da ist.

17.

Today equator celebration with Neptune and baptism. Much harmless larking about. Played music at various times with the merry widow Ohnesorg and a young, overly made-up Chilean woman. Because attendance at the party on the 19th is obligatory. I also have to deliver a relativity lecture to the officers. Splendid isolation is crumbling. But the voyage won't last forever, either. Heat 22 Réamur in the shade.[23] Thoroughly sweaty at night. I have become convinced that $\left(R_{ik} - \frac{1}{4} g_{ik} R\right) = T_{ik_{el}}$ is not the right thing. Conviction about the impossibility of field theory in its meaning so far is intensifying.[24] What a pity that Margot isn't here.

11

Allein ist schön, aber nicht
alleins unter viel fremden Affen.

19.

Vorgestern Aequatorfest im 1., gestern
im 2. Klasse mitgemacht. Bei ersteren
Argentiner schlecht abgeschnitten.
Reiche Klasse. Blasiert und dabei
Kinder. Bei letzteren naive
dankbare Menschen. Kapitän
schöne Witze (Schrot im Urin,
Patient mit den von oben nach unten wandernden
Schmerzen). Heute Besuch
des Maschinen und Kesselraums
Grosser Eindruck. Abends
Konzert in 1. Klasse. Ich spielte
im Quartett Mozarts Nachtmusik
u. dann Beethovens Romanze
in f-dur. Argentiner unsagbar
blöde Produktionen. Und für
mich – was Geistigkeit in d son-

Solitude is nice, but not alone among many foreign apes.

19.

Day before yesterday, equator celebration in first class; yesterday in second class. In the former, the Argentines cut a poor figure. Rich class. Blasé, but also childish. In the latter naïve, grateful people. Captain, good jokes (pellets in urine; patient with pain traveling from top down). Today, visit of the engine and boiler rooms. Great impression. In the evening, concert in 1st class. I played in a quartet Mozart's *Nachtmusik* and then Beethoven's Romance in F major.[25] Argentines unspeakably stupid creatures. I'm done with them, at least,—as far as intellect and

12

stigen Gehalt anlangt erledigt, wenigstens
die Mdr F. Jesinghaus zeigte mir
populäre urgent Musik, von Inkas
herrührend. Naturhaft und gross.
Da muss herrliches untergegangen
sein an diesem Volk. Necke mich
viel mit Panterkatze, die mich
unausgesetzt ergründet. Sie ist
amüsant in ihrer ernsten und
impertinenten Art. Jüdin von
russischem Typus

Ich spiele jeden Tag mit einem
H. Holländer, Frau Ahlesorg und
dem Primgeiger der Schiffskapelle
Quartett, Mozart und Schubert
in meiner Kajüte. Man schwitzt
arg dabei, aber man ist beglückt.

Idee zur ~~Begründung~~ Erklärung
der Kohärenz von nach verschiedenen
Richtungen emittierter Strahlung. (18.III.)

other substance are concerned—M[embers] [of the] I[dle] R[ich] C[lass].[26] Jesinghaus showed me popular Argentine music, originating from the Incas. Naturalistic and grand. Glorious things must have perished with that nation. I tease panther cat a lot, who is constantly probing me. She is amusing in her serious and impertinent way; Jewess of Russian type.

I play quartets, Mozart and Schubert, every day with a Mr. Holländer, Mrs. Ohnesorg, and the first violinist of the ship's band in my berth. One sweats terribly doing so, but one is delighted.

Idea about explanation for the coherence of radiation emitted in different directions.[27] (18 March).

13

22

Gestern Rio Rabbiner u. sonst
einer sowie einige Ing. und Medi-
ziner holten mich am Schiff
ab. Zuvor $5\frac{1}{2}$–7 Einfahrt in den
Hafen. Himmel bedeckt und
schwacher Regen, aber doch maje-
stätischer Eindruck der bizarren
Riesenfelsen. – Meine Begleitung
sehr gemütlich und angenehm.
Botanischer Garten wie überhaupt
Pflanzenwelt übertrifft die Träume
von 1001 Nacht. Alles lebt und
wächst sozusagen unter den Augen.
Köstlich ist das Völkergemisch
in den Strassen. Portugiesisch-indianisch
-Negerisch mit allen Übergängen
Pflanzenartig triebhaft, hitzegedämpft.
Wundervolles Erlebnis. Eine unbeschreib-
liche Fülle von Eindrücken in

22.

Yesterday Rio, Rabbi and someone else as well as some engineers and medical doctors picked me up at the ship.[28] Earlier 5:30–7 a.m. entry into harbor. Sky overcast and light rain, but nevertheless a majestic impression of the bizarre giant cliffs—My company, very cozy and pleasant. Botanical garden, indeed the plant world in general, surpasses the dreams of 1001 Nights.[29] Everything lives and thrives, so to speak, under one's very eyes. The miscellany of peoples in the streets is delightful. Portuguese–Indian–Negro, and everything in between, plant-like and instinctive, subdued by heat. Wonderful experience. An indescribable abundance of impressions in a

14. Postcard sent by Einstein to Elsa and Margot Einstein depicting a vista of the Botafogo neighborhood in Rio de Janeiro and including a note in Einstein's handwriting: "Keep the postcard because it's pretty," 5 May 1925 (Courtesy Albert Einstein Archives).

wenig Stunden.

27.

24. Montevideo mittags. Journalisten und andere Juden verschiedener Sorte, unter anderen Nierenstein, Sekretär der Universität. Dieser ist ein resignierter, guter Mensch, aber die anderen alle mehr oder weniger unsauber.

Schiff sollte 12 Uhr in Buenos eintreffen, blieb aber gegen 11 Uhr stecken, sodass erst 2 ½ Nachts Ankunft. Bin halbtot von dem unappetitlichen Gesindel. Frau Jerusalem und alle [illegible] stehen mir bei. 7 Uhr Morgens geht der Tanz auf dem Schiff wieder los. 8 ½ gingen wir ans Land. Nierenst. hilft. Fahren zu Wassermanns. Endlich Ruhe, ganz abgehetzt. Gütige und frohe Haus

few hours.

27.

24. Montevideo at noon. Journalists and other Jews of various sorts, among others Nierenstein, secretary of the university. He is a good person, resigned to his fate, but the others are all more or less sordid.[30]

Ship was supposed to arrive in Buenos Aires at 5 p.m., but ran aground about 11 a.m., so arrival only at 2:30 at night. Am half-dead from the unsavory riff-raff. Mrs. Jerusalem and all the stewards are standing by me. 7 a.m., the dance begins again on board ship. 8:30 we go ashore.[31] Nierenstein assists. Drove to Wassermanns.[32] Peace at last; completely exhausted. Kind and cheerful

15

Frau. Nimmt alle Chikanen
einer „freiwilligen Sekretärin"
auf sich zusammen mit Frau
Jerusalem. Nachmittags Kreis
der Freundinnen („Soldaten") da
sowie deutscher Gesandter.
Abends Besuch bei Rektor und
Dekan. Schlichte und freundliche
Menschen ohne Pose, aber auch
ohne das Gefühl einer Mission.
Nüchtern, aber die und andere
richtige Republikaner, in manchem
an Schweizer erinnernd.

Stadt komfortabel und
langweilig. Menschen zart,
Rehaugen, Grazie, aber schablonenhaft.
Luxus, Oberflächlichkeit.

Am 26. Menge Journalisten
und Photographen. Um 12 Uhr
Fahrt in Stadt und zum Markt.

housewife. Assumes all the chicanery of a "volunteer secretary" together with Mrs. Jerusalem. Afternoon, ladies' circle ("soldiers") there, as well as German ambassador.[33]

In the evening, visit to rector and dean.[34] Unassuming and friendly people without pretentiousness, but also without any sense of a mission. Sober, but they and others are genuine republicans, in some ways reminiscent of the Swiss.

City comfortable and boring. People delicate, doe-eyed, graceful, but clichéd. Luxury, superficiality.

On 26th, a multitude of journalists and photographers. Around 12 p.m., drive into town and to market.[35]

16

Südlich gemildertes New York. Nachmittags Dekan von La Plata ziemliches Repräsentiermenscherl mit analogem Frauchen und Deputation der Juden. Letztere wollen mich in Massenversammlung „feiern". Da ich aber Nase pleng von New York, lehne ich entschieden ab. Nur Zusammenkunft mit Deputationen bleibt übrig. Abends Familie. Frau Jerus, spielt sich auf mit sehr viel Witz und Geist, aber mit forcierter Lustigkeit.

Heute Vormittag ruhig zuhause. Nachmittag Empfang in der Universität mit einleitendem Vortrag.—

Langer Saal, Cylindergewölbe. Schwungvolle Reden, ich französisches Gestöpsel bei Unruhe. Kulturlose Angelegenheit.

New York attenuated by the South. In afternoon, dean from La Plata, dainty homunculous with analogous little wife[36] and deputation of the Jews. The latter want to "celebrate" me in a mass gathering. But as I'm fed up[37] with New York, I resolutely decline.[38] Only meeting with deputations remains. Evening with family. Mrs. Jerusalem performs with a great deal of wit and intellect, but with forced hilarity.

[27.] This morning quietly at home. Afternoon reception at the university with introductory lecture.—[39]

Long hall, barrel vault. Rousing speeches; I, mumbling in French amidst commotion. Philistine affair.

15. Einstein's inaugural lecture at the Colegio Nacional high school, Buenos Aires, 27 March 1925 (Courtesy Archivo General de la Nación, Buenos Aires).

17

28. Erste Vorlesung im überfüllten Saal bei Siedehitze. Jugend ist immer erfreulich, weil interessiert für die Dinge. Sympathischer Unterrichts-Minister ist auch da. Viel belanglose Besuche, aber gut zum Aushalten.

Am Abend des 27. Kleine Gesellschaft bei Hirsch. Er gewaltiger Willensmensch, sie schöne Jüdin [mit eigenmächtigende übersetzt] Typus Alice. Haus mit prächtigen Bildern und Orgel. Die Sucht, das Schöne für sich haben zu wollen [bezw. durch Kauf an sich reissen zu wollen], ist der erste Schritt vom Barbaren zum feineren Menschen. Vergleichbar dem Kinde, dem es nicht genügt, sich am Anblick des Schmetterlings zu freuen, sondern das nach ihm tappen oder ihn gar ins Maul nehmen muss, um die ihm eigen-

28. First lecture in overfilled hall in seething heat. The young are always pleasant because they are interested in the topics. Likable education minister is also there.[40] Many insignificant visits, but tolerable enough.

In evening of the 27th, small party at Hirsch's. He, a powerful, willful person; she, a beautiful Jewess, an Alice-type but triumphant in nature.[41] House with magnificent pictures and organ. The craving to take possession of beauty, or buy it up and grab it, is the first step from the barbarian to the refined person. Comparable to a child, for whom it is not enough to be pleased at the sight of a butterfly, having rather to grope after it or even put it into its mouth in order to get the

tümliche Befriedigung zu erlangen.

29. Regensonntag in glückseliger Ruhe allein auf meinem Zimmer am Vormittag. Nachmittag Spaziergang mit H. Wassermann. Man braucht ziemlich viel äussere Unruhe, um in der Ruhe Seligkeit finden zu können. Meine Vorbereitung hiefür in der letzten halben Woche war reichlich genügend. 30. Regenso Montag 30. Nachmittags Vorlesung mit Diskussion. Abends bei Robert mit Hofer; wie sind wir alt geworden! 12–1½ Besuch bei Prensa. Ungeheurer Aufwand von Papier und Menschen und Technik für ?. Traurige Zeit kommt einem zu Bewusstsein.

31. Besuch der Redaktion von „Das Volk" und im Judenviertel sowie mehreren „Schulen" und Waisenhaus. Die Tragik des jüdischen Volkes; es verliert mit den Läusen die Seele. Ist es nicht

satisfaction peculiar to a child.

29. Rainy Sunday in blissful peace alone in my room in the morning. One needs quite a lot of external disquiet in order to find one's bliss when it is quiet. My preparation for that in the past half-week was copious enough. Afternoon walk with Mr. Wassermann. 30. Rainy Monday 30. In afternoon, lecture with discussion. Evening at Robert's together with Hofer; how much we have aged![42] 12–1:30 p.m. visit to *Prensa*. Immense expenditure on paper and people and technology for? One realizes how sad the times are.[43]

31. Visit to the editorial office of *Dos Volk* and to the Jewish quarter as well as many "shuls" and the orphanage.[44] The tragedy of the Jewish people: It loses its soul along with its lice. Is it

19

mit andern Völkern auch so ähnlich? Ich glaube doch nicht so ausgesprochen.

Nachmittags bei Hofer auf seinem Landgut im „el Tigre". Hat eine Tochter von echt Schweizer Art und lebt sehr beschaulich.

1 IV. Vormittags Flug über Stadt mit Junker-Wasserflugzeug mit Frau Wassermann. Erhabener Eindruck, besonders beim Auffliegen.

Nachmittags Besuch bei Minister und Präsident. Besuch des Völkermuseums, Vortrag und Abendessen bei Lugones. Das reicht.

2 IV. Besuch in La Plata. Hübsche, stille, italienisch anmutende Stadt mit prächtigen Universitäts-

similar with other peoples too? I do not believe quite as markedly.

In the afternoon, with Hofer on his country estate in “El Tigre.” Has a daughter of genuine Swiss nature and lives a placid existence.

1 April. In the morning, flight over city in Junker[s] hydroplane with Mrs. Wassermann.[45] Sublime impression, especially during ascent.

In the afternoon, visits to ministers and president. Visit to the ethnological museum, lecture, and dinner at Lugones’s.[46] That’s more than enough.

2 April. Visit to La Plata. Pretty, quiet, Italian-like city, with magnificent university

Crítica

EL PROFESOR EINSTEIN VOLO HOY SOBRE LA CIUDAD

REALIZO UN VUELO EN AEROPLANO, ACOMPAÑANDO AL PILOTO ALEMAN GRUNDKE

El piloto Grundke que cargó con la responsabilidad de llevar por el aire al sabio; Einstein y el mecánico, antes de partir

Al regresar del vuelo, Einstein desciende del hidroplano con la dama que lo acompañó en el vuelo

Allí va Einstein, sobre la ciudad, desafiando un peligro . . . relativo

16. Einstein’s flight over Buenos Aires, *Crítica*, 1 April 1925 (Courtesy Biblioteca Nacional Mariano Moreno, Buenos Aires).

20

gebäuden, die nach nordamerikanischer Art eingerichtet sind. Besuch des sehr interessanten naturhistorischen Museums. Einweihungsfeier des Semesters mit sehr langer Rede und musikalischen Vorträgen.

3 IV. Essen von Rektor. Vortrag

4. Abends Robert mit Dreyfus. Letzterer gescheit und dabei gemütlich. Macht bürgerlich-pfiffig-gutmütigen Eindruck. Vorher nachmittags Vorlesung in philosoph. Fakultät über anschaul. Vorst. des sphärischen Raumes.

5. Mit Wassermanns Autofahrt in deren Gut Lavajol.

6. Morgens Besuch mit hiesigem Physiologen bei Augenarzt und Börsenmakler Fortin. Zeigte Experimente über subjekt. Ersch.

buildings which are furnished in the North American style. Visit to the very interesting natural history museum. Inaugural ceremony of the semester with a very long speech and musical presentations.[47]

3 April. Meal with rector. Lecture.[48]

4. Evening, Robert with Dreyfus. The latter clever and yet pleasantly jovial. Makes bourgeois, sharp, good-natured impression. Beforehand, afternoon lecture in Faculty of Philosophy on the intuitive representation of spherical space.[49]

5. With Wassermans, drive to their estate Lavajol.[50]

6. In the morning, with local physiologist, visit to the ophthalmologist and stockbroker Fortin. Demonstrated experiments on subjective phenomena

21

auf der Retina bei Verwendung
von intens. monochromatischem
Licht. Nachmittags Vortrag. Abends
grosse Versammlung der Zionisten
im Theater wegen Univ. Jer. Spanier
mit eleganten Pablos traten auf.
Ich kurze Ansprache. Mossinsohn
auf jiddisch in volkstümlicher
Weise. Das jiddisch ist von merk-
würdiger Innigkeit des Ausdrucks.
7. Besuch in Klinik des Rektors Arce.
Gute Einrichtungen. Tüchtiger
Mann. Sticht sehr ab von Umgebung.
8.9.10. Landaufenthalt in Lavayol
Schöpfte neue Kräfte. Original
Don Pablo Ameisen. Schönes Wetter
wunderbare Ruhe. Prächtige Idee
für neue Theorie des Zus. Grav.
Elektrizität.
11.12. Abends Reise im Extrawagen

on the retina through application of intense monochromatic light. Afternoon lecture. In the evening, large assembly of Zionists at the theater regarding the Jerusalem university. Spaniards appeared on the stage with elegant pathos. I, short speech. Mossinson in Yiddish in folksy manner.[51] Yiddish has a remarkable intimacy of expression.

7. Visit to clinic of rector Arce.[52] Well equipped. Capable man. Stands out very much against his surroundings.

8. 9. 10. Country sojourn at Lavajol. Replenished new energy. Original Don Pablo ants. Fine weather, wonderful quietude. Splendid idea for a new theory of the connection between gravitation and electricity.[53]

11. Evening trip in special railway car

17. Einstein's stopover in Sunchales, Santa Fe province, en route to Córdoba, 12 April 1925 (Courtesy Centro Marc Turkow, Asociación Mutual Israelita Argentina, Buenos Aires).

22

Loyarte
mit Ing. Butty, Phil. Dekan Alberini
und Ing. Dekan nach Cordoba.
12. Sonnt. Autofahrt ins altes, malerisch-
kärglich bewachsenes Granitgebirge
Abends sehr langweiliges Essen
v. Regierung.
wunderv. Saal dar.
13. Festsitzung und Vortrag in Universität
Mittagessen neben neuem Gouverneur
der Provinz, einem sehr feinen, interessanten Menschen. Sonst nur ermüdende
Fülle von Spaniern, Journalisten
und Juden drollige hebräische Anekdote
von zitternder Jungfrau. Abends
Schlaf in Extrawagen
14. Rückfahrt nach Buenos. Ausser
Alberini gew. Menschen. Glücklich
bei Ankunft. Bin schrecklich
menschenmüde, Der Gedanke,
noch so lange herumziehen
zu müssen, lastet schwer auf mir.

with engineer Buty, Loyarte, dean of Philosophy Alberini, and dean of Engineering to Cordoba.[54]

12. Sunday. Car ride into ancient, picturesque, sparsely vegetated granite mountains. Evening, very boring meal [hosted] by government.[55]

13. Festive session and lecture in wonderful hall of the university. Midday meal beside new governor of the province, a very refined, interesting person. Otherwise only tiresome plethora of Spaniards, journalists, and Jews; droll Hebrew address by trembling maiden.[56] In the evening, sleep in the special railway car.

14. Return trip to Buenos. Apart from Alberini, average people.[57] Happy to arrive. I'm terribly weary of people. The thought of still having to gallivant about for so long weighs heavily on me.

18. Einstein arriving in Córdoba with university faculty including Law Professor Guillermo Rothe and rector Léon S. Morra, 12 April 1925 (Courtesy Archivo General de la Nación, Buenos Aires).

23

Gesamteindruck lackierte Indianer
skeptisch-zynisch ohne Kulturliebe, im
Ochsenfett verkommen. Cordoba zeigt
Reste echter Kultur mit Liebe
zum Boden und Sinn für das
Hohe. Wunderbare Kathedrale. Häuser
fein proportioniert (altspanisch) ohne blöden Schmuck
Dafür aber Pfaffenherrschaft. Ist ~~also~~
immer noch besser als suffisante
Zivilisation ohne Kultur

16. 10h Sitzung der zionistischen Exekutive
im dortigen Bureau. Sehr feierlich
Als mir Kuriositäten gezeigt wurden
fand sich unter Wand-Photographie
ungeheurer Dreck. Hoffentlich nicht
als Symbol zu wirken.

Nachmittags Sitzung der Akademie
Wurde auswärtiges Mitglied. Man stellte
sehr dumme (wissenschaftliche) Fragen an mich, sodass
es schwer war, ernst zu bleiben.

Overall impression, lacquered Indians, skeptically cynical without any love of culture, debauched in beef tallow. Cordoba exhibits vestiges of genuine culture with love of the soil and a sense for the sublime. Wonderful cathedral.[58] Buildings finely proportioned (old Spanish) without daft ornamentation. But on the other hand, clerical rule. It's still better than a smug civilization without any culture.[59]

16. 10 a.m. session of the Zionist executive in the office there. Very festive.[60] When I'm shown curios, tremendous filth was revealed under a photograph on the wall. I hope it's not to be taken as a symbol.

In the afternoon, session of the Academy. Became foreign member. Very stupid scientific questions were posed to me, so it was difficult to remain serious.[61]

19. Einstein at the University of Córdoba, 13 April 1925 (Courtesy Archivo General de la Nación, Buenos Aires).

24

17. Früh vom Maler photographiert. Nachmittags vorletzte Vorlesung. Abends Empfang bei der deutschen Botschaft. Lauter Hiesige, keine Deutsche, denn der Gesandte scheint es nicht gewagt zu haben, letztere zu mir einzuladen. Drollige Gesellschaft, diese Deutschen. Ich bin ihnen eine stinkende Blume und sie stecken mich doch immer wieder ins Knopfloch.

18. Nachmittags Privatvortrag in Wassermanns Haus für Regiment ohne Panterkatze. Letztere protest wegen Vernachlässigung. Abends Societe Hebraika. Vortrag über Wesen des Zionismus und Grösse der Atome.

19 Sonntag. Ausflug mit W. nach Lavajolle. Abends bei Zaslavski mit Robert. Dann Empfang durch Vertretung jüdischer Kreise im Hotel. Reden von Gesang und Mossinson.

17. Photographed early by painter. Afternoon, next to last lecture. Evening reception at the German Embassy. Nothing but locals, no Germans; for the ambassador seems not to have dared to invite the latter to see me.[62] Funny bunch, these Germans. To them I'm a stinking flower, yet they stick me back into their button hole over and over.

18. In the afternoon, private lecture at Wassermanns' home for regiment, without panther cat. The latter *broges* owing to neglect. In the evening, Hebrew Society. Lecture on the nature of Zionism and size of atoms.[63]

19. Sunday. Excursion with W[assermann] to Lavajolle. In the evening, at Zaslavski's with Robert. Then reception by representation of Jewish associations in hotel. Speeches by Gesang and Mossinson.[64]

25

20. Letzter wissenschaftlicher Vortrag mit Begeisterung des Publ.

21. Vormittags sehr geschmackloser Empfang im jüdischen Spital. Ich habe die Leute heruntergeputzt. Mittags Frühstück der engeren Kollegen im Klubhaus Tigre. Abends Widmungsgedichte für Photographien:

Frau Wassermann:

Herrgott was sie alles kann
Die signora Wassermann
Durch das Telefon regieren
Und der Haute Volée servieren
Auch verbreiten liebe Tugend
Bei der – ach! – schon ältren Jugend
Warm Gemüt und hell Verstand
Dankend reich ich ihr die Hand.

20. Last scientific lecture to the raptures of the audience.[65]

21. In the morning, very tasteless reception at the Jewish hospital. I cut those people down to size. Noon, breakfast hosted by more closely acquainted colleagues in the Tigre clubhouse.[66] Evening dedication poems for photographs:

Mrs. Wassermann:[67]

God Almighty, what many things she can do,
That signora Wassermann.
Govern by telephone
And cater to high fliers
Also spread austere virtue
Among the—alas!—now older youth.
Warm temperament and clear mind
Thankfully I extend her my hand.

26

Prof. Nierenstein

Wohl Geführt von Ihrer Hand
Tappt' ich tapfer durch dies Land
Wer weiss so, was sich gebührt,
Dass nur keiner mopsig wird?
Allen sieht ins Herz hinein
Herr Professor Nierenstein.
Dank an seine Seele mild
Künd' ihm dieses Tabóo-Bild

Panterkatze:

Dieses für die Panterkatze
Zornig sich verkrochen hat se
In den Dschungel bös und wild
Also kriegt sie dieses Bild.

22. Offizielles Frühstück für Gross-
köpfete der Wissensch. u. Politik
(Gallardo, Rektor Dekane Gesandte etc.)
Abends Studenten Guitarre u. Gesang
ich zum Schluss mit Geige

Professor Nierenstein[68]
Well-guided by your hand
I trotted bravely through this land.
Who knows what is proper,
To avoid anybody getting broges?[69]
He sees into everyone's heart,
Professor Nierenstein
My thanks to his mild soul
Conveyed by this portrait of a sabio.[70]
Panther cat:[71]
This one's for the panther cat
She has slunk away in fury
Into the jungle, angry and wild
Thus this is the portrait she receives.

22. Official breakfast for bigwigs of science and politics (Gallardo, rector, deans, diplomats, etc.). In the evening, students, guitar, and singing; at the end I play the violin.[72]

27

23. Viel Einzelheiten. Scharf schenkt mir Ölbild. Abend Abreise

24. 8 Uhr morgens Ankunft Montevideo. Beschliesse Krankheit. Übersicht über M. vom Versicherungs-Gebäude. Einquartiert bei russisch jüd. Familie Rosenblatt. Besuch des deutsch. Gesandten. Ras Fereiden Bummel Feiner schwarzer nervöser Kerl. Spricht schlecht französisch, noch schlechter als ich. Hatte Scheu vor mir wie die meisten.

In Uruguay fand ich eine ächte Herzlichkeit wie selten in meinem Leben. Ich fand dort Liebe zum eigenen Boden ohne irgend welchen Grössenwahn. Nach der Ankunft mit Rosenblatt und

23. Many little details. Scharf gives me an oil painting. Evening departure.[73]

24. 7 a.m., arrival Montevideo. Decide that I'm ill. Skyline view of M[ontevideo] from insurance building. Quartered with Russian–Jewish Rosenblatt family. Visit from German ambassador. Stroll with Ras Fereida, decent, black, nervous fellow.[74] Speaks French badly, even worse than I. Was shy in my company, as are most.

In Uruguay I was met with genuine cordiality like seldom before in my life. There I encountered love of one's own soil without any kind of megalomania. After arrival with Rosenblatt and

20. Einstein disembarking from ship upon arrival in Montevideo, 24 April 1925 (Courtesy Leo Baeck Institute, New York).

28

Söhnen vom Versicherungs-Haus aus prachtvoller Blick auf Stadt und Hafen. Die Familie sehr herzlich und treuherzig. Er und sie nur jiddisch, die Kinder nur französisch.

Prof. Ing Maggiolo sehr lieber, feiner Mensch, leise und in sich gekehrt, gar nicht amerikanisch. Ing Castro jüngerer auch netter Mensch mit reizendem rotbackigem Söhnchen.

25. Erste Vorlesung mit feierlichem Empfang. Abends mit Maggiolo und Rossenblatts, deutschem Gesandten Traviata von italienischer Truppe. Recht hübsch.

26. (Sonntag) Küsten-Spaziergang mit Bürgermeister. Sehr hübsch, mit Sonnen-Untergang. Geschmackvolles Strandhotel wurde mir gezeigt, von einem Einheimischen gebaut. Abends Lohengrin mir zuliebe

sons, splendid view onto city and harbor from insurance building. The family very welcoming and trusting. He and she only Yiddish, the children only French.[75]

Prof. Eng. Maggiolo very kind, decent person, quiet and introverted, not at all American. Eng. Castro younger, also nice person with charming red-cheeked little son.[76]

25. First lecture with ceremonial reception.[77] Evening with Maggiolo and Rossenblatts and German ambassador. Traviata by Italian company. Pretty nice.

26. (Sunday) Walk along the coast with mayor. Very pretty, with sunset. Tasteful beach hotel was shown to me, built by a native. In the evening, Lohengrin

21. Einstein and Amadeo Geille Castro, Montevideo, 24 April 1925 (Courtesy Leo Baeck Institute, New York).

29

gespielt. Schwankte zwischen gut und komisch. Liegt nicht nur an der Truppe! Zwei Studenten halten immer Wache, dass kein Unberufener zu mir kommt. Habe einen rührenden Diener zugeteilt bekommen, mit dem ich mich nur mit den Händen verständigen konnte.

Uruguay glückliches Ländchen, nicht nur liebliche Natur mit angenehmem feuchtwarmem Klima sondern auch mit vorbildlichen sozialen Einrichtungen. (Mutter- & Kinder-Schutz, Versorgung alter Leute und unehelicher Kinder, 8-Stundentag, Ruhetag). Sehr liberal Staat von Kirche ganz getrennt. Verfassung der schweizerischen einigermassen ähnlich. Montevideo

played in my honor. Fluctuated between good and comical. Not just because of the performers![78] Two students keep constant watch, so that no unauthorized person approach me. A heart-warming servant was assigned to me, with whom I could only communicate through gestures.

Uruguay, happy little country, is not only charming in nature with pleasantly warm humid climate, but also with model social institutions. (Mother & child protection, care for old people and illegitimate children, 8-hour workday, day of rest). Very liberal, state completely separated from church. Constitution somewhat similar to the Swiss one.[79] Montevideo

architektonisch hübsch im Kolonialstil.
27. Morgens mit Senats-Präsident
in Fabrik zur Bearbeitung des einheimischen,
sehr schönen und mannigfaltigen
Marmors. Sehr gescheiter aber gerissener
jüngerer Mann, der Lugones in
der Kommission d. Coop. Int. des V. B.
vertreten soll. Maggiolo und Castro
sowie einige andere waren auch
dabei. Dann Besuch des neuen,
fast vollendeten Regierungsgebäudes.
Sehr geschmackvoll in Hochrenaissance
vornehmen, von italienisch-schwei-
zerischen Architekten ausgeführt.
Nachmittags Besuch bei Präsidenten
der Republik und Unterrichtsminister
(Auch bei Schwager-Konsul Geyer, der in Aarau Schüler war)
Ersterer interessanter Kopf letzterer
in sehr schönem altspanischem
Hause. Dann Vorlesung. Abends
Familie Rossenblatt. Drei Söhne, zwei

architecturally pretty, in colonial style.

27. In the morning with Senate president in factory for treatment of the indigenous, very beautiful and varied marble. Very smart but shrewd younger man, who is supposed to represent Lugones on the Committee on Intellectual Cooperation of the League of Nations. Maggiolo and Castro as well as some others were also present. Then visit to the new, almost completed government building. Very tasteful interior in high Renaissance, carried out by an Italian–Swiss architect.[80] In the afternoon, visit with the president of the Republic and education minister. (Also to the Swiss consul Guyer, who was a schoolboy in Aarau.) The former an interesting mind, the latter in a very beautiful old Spanish building. Then lecture.[81] In the evening, Rossenblatt family. Three sons, two

31

verheiratet und eine nicht hübsche
aber gutherzige verlobte Tochter.
28. 6 Uhr Empfang der deutschen
Kolonie. Gemütlich und angenehm
mit Kaffee-Begleitung. Wahrschein-
lich waren nur die liberalsten erschienen
Abends feierliches Bankett der Juden
Völkerbundskommission für Immigration von Kriegs-Vertriebenen
war dabei. Sass neben interessantem
Engländer (Nansens Mitarbeiter)
Latzki (Russe, in Berlin wohnend)
war auch dabei. Erhielt von
ihm Empfehlungsbrief an Gallardo.
29. Empfang in ~~Polyt~~ Ing. Schule
~~Dipl~~ Gedenkmedaille der Studenten.
Letzte Vorlesung. Abends grosser
Empfang bei deutschem Gesandten,
bei dem nur Uruguayische Politiker
und Gelehrte waren.

married, and one not pretty but kind-hearted engaged daughter.

28. 6 p.m. reception hosted by the German colony. Cozy and pleasant accompanied by coffee. Probably only the most liberal showed up. In the evening, formal banquet put on by the Jews. League of Nations Commission for Immigration of War Refugees was in attendance. Sat next to interesting Englishman (Nansen's coworker). Latzki (Russian, residing in Berlin) was also there. Received from me a letter of recommendation to Gallardo.[82]

29. Reception at School of Engineering. Memorial medal from students. Last lecture. In the evening, major reception by the German ambassador attended only by Uruguayan politicians and scholars.[83]

22. Einstein visiting the Montevideo College of Engineering, with Amadeo Geille Castro and dean Donato Gaminara, 29 April 1925 (Courtesy Leo Baeck Institute, New York).

32

30. Morgens Kino Südpol-Expedition, austral. Archipel und ein hübscher Shapley-Films (der falsche Pastor) eigens für mich von Glücksmann aufgeführt. Nachmittags wundervolle Segelpartie. 6 Uhr Empfang im Ingenieurverein. Abends 9 Uhr grosses Bankett gegeben von Regierung und Universität. Ich sass neben Präsident und einem Minister (Wacht am Rhein statt deutsche Hymne gespielt; Deutsche Ges. und ich schmunzeln) und unterhielt mich vortrefflich. Die Menschen waren rührend und ohne Ceremoniell. Aber ohne Smoking gehts nicht

1. VI. Bemerkung: das über Montevideo ist aus dem Gedächtnis am Dampfer geschrieben. In Wirklichkeit war es viel mehr und viel bunter, sodass ich bei aller Liebe manchmal kaum mehr japsen konnte. Es war aber

30. In the morning, cinema. South Pole expedition, Australian archipelago, and a nice Shapley [Chaplin] film (The Fake Pastor), shown especially for me by Glücksmann.[84] In the afternoon, wonderful sailing party. 6 p.m. reception at the Engineers' Association. 9 p.m., grand banquet thrown by government and university. I sat next to president and one minister and was excellently entertained. Wacht am Rhein played instead of German anthem![85] German ambassador and I grinned. The people were touching and did not stand on ceremony. But nothing goes without a tuxedo.

Comment: the text about Montevideo is written from memory on the steamship.[86] There was much more in reality and it was much more colorful, to the point that at times I could hardly catch a breath for all that love. But it was

33

viel menschlicher und erfreulicher als in Buenos Aires, wozu natürlich die kleineren Dimensionen des Landes und der Stadt beitrugen. Diese Leute erinnern eben an Schweizer und Holländer. Bescheiden und natürlich. Hol' der Teufel die grossen Staaten mit ihrem Fimmel. Ich würde sie alle in kleinere zerschneiden, wenn ich die Macht dazu hätte.

1 V. Alle Arbeit ruht und keine Autos dürfen fahren. Ich werde mit Stadt-Auto zur Bahn gebracht und mit Hafendampfer samt der zahlreichen Begleitung aufs Schiff gebracht. Valdivia, französisch. Sehr dreckig und klein, aber freundliche Mannschaft und gemütlich. Nur

much more human and enjoyable than in Buenos Aires, to which the smaller dimensions of the country and the city contributed, of course. The people simply remind one of the Swiss and Dutch. Modest and natural. The devil take large countries with their fads! I would cut them all up into smaller ones, if I had the power to do so.

1 May. All work rests and no cars may drive. I am brought in a city automobile to the railway and taken by harbor steamboat, together with all my numerous company, on board the ship. Valdivia. French, very filthy and small, but friendly crew and cozy.[87] But

23. Einstein departing Montevideo, 1 May 1925 (Courtesy Archivo *El País*, Montevideo).

34

vor dem Abtreten graust mir.
3 Tage kann man's aber aushalten.
Meine Nerven sind abgespannt. Ich
gäbe was drum, wenn ich in Rio
nicht noch einmal aufs Trapez
müsste. Aber man muss aushalten.
2 V. Der Schiffsarzt hat mir ein
Buch von Le Bon mit Aphorismen
zur Politik und Soziologie gegeben.
Geistreich, aber nicht frei von gewissen
Vorurteilen, insbesondere das kommunis-
tische Problem betreffend. Er raisonniert
wie die Liberalen von 1850. Auch ist
er nicht frei von Militarismus.
Es wird ziemlich heiss. Dazu das
ziemlich schwere und schlecht
gekochte Essen. Man schläft schlecht. Gesellschaft am
„Honoratiorentisch" recht gemütlich,
besonders Kapitän. Sind viel angenehmer
als Deutsche, viel schlichter und

I shudder at the latrines. But one can put up with it for three days. My nerves are strained. I'd give a lot not to have to climb back onto the trapeze in Rio. But one must stick it out.

2 May. The ship's doctor gave me a book by Le Bon with aphorisms on politics and sociology. Witty, but not free of certain prejudices, particularly as regards the Communist problem. He reasons like the liberals of 1850. Nor is he free of militarism.[88] It's getting quite hot. Add to that the quite heavy and badly cooked food. One sleeps poorly. Company at the "dignitaries' table" is quite good-natured, particularly the captain. They are much more pleasant than Germans, much more unpretentious and

35

natürlicher. Dabei von einem gewissen
Feingefühl, nicht zudringlich. All
meine wissenschaftlichen Ideen, die
ich in Argentinien ausdachte, erwiesen
sich als unbrauchbar. Das Wetter
ist schlecht bis mittelmässig.
3. Wetter etwas besser. Viel Wind, aber
wenig erfrischend. Die Ruhe aber
thut wohl. Morgen abend ist die
Herrlichkeit vorbei, und ich muss
ein letztesmal aufs Trapez. Diese
paar Tage Affenkomödie werde
ich mit Gottes Hilfe noch aushalten.
Dann kommt dafür eine schöne
lange Heimreise. Ich kann mir
ein regelmässiges, stilles Leben kaum
mehr vorstellen, soviel Unruhe und
Wechsel liegen hinter mir. Wie werden
sich unsere Papierhelden freuen, dass
sie dem Michel den Hindenburg auf-

natural. At the same time, they have a certain sensitivity, not intrusive. All my scientific ideas that I thought up in Argentina prove to be useless.[89] The weather ranges from bad to middling.

3. Weather somewhat better. Much wind but scarcely refreshing. The tranquility does feel good, though. Tomorrow evening this splendor is over, and I must climb onto the trapeze one last time. With God's help, I shall survive these couple of days of buffoonery. In return for that, there'll then be a very nice long trip homeward. I can scarcely imagine a regular, quiet life anymore; that's how much hustle and bustle lies behind me. How our paper heroes will enjoy having persuaded the plain honest German to vote for Hindenburg.[90]

36

geschwatzt haben. Dem Deutschen Gesandten in Montevideo war's peinlich, und die Uruguayer machten sich über die Deutschen lustig: Die Nation, der man mit dem Stock die Klugheit ausgetrieben hat.
4. Ankunft in Rio bei Sonnenuntergang und prächtigem Wetter. Granitfels-Inseln von phantastischen Formen sind vorgelagert. Feuchtigkeit gibt geheimnisvolle Wirkung. Schiffsarzt erzählt von zwei Erlebnissen über Telepathie (Er träumt, vom Bauer vom Lande abgeholt zu werden zu 17jähr. Tochter wegen Geschwulst in Achselhöhle, die aufgeschnitten werden muss. Hat sich nächsten Tag genau bewahrheitet.) Er erzählt von zwei anderen Fällen, die ihm selbst passiert sein sollen. Gebildeter und nüchterner Mann. Zeigt

It was embarrassing for the German ambassador in Montevideo; and the Uruguayans were poking fun at the Germans: The nation that had its reason beaten out of it with a stick.

4. Arrival in Rio at sunset and in splendid weather. Fantastically shaped granite-rock islands in foreground.[91] Humidity lends a mysterious effect. Ship's doctor tells of two experiences with telepathy. (He dreamed of being called for by a country farmer about a growth in the armpit of a 17-year old daughter, that had to be sliced open. The next day, precisely that really happened.) He tells about two other cases that supposedly happened to him. Educated and sensible man.

37

nur auch Notiz über Pariser Experimente über Bakteriophagen, die ich aber bestimmt für falsch interpretiert halte.

Von Hotelmenschen im Hafen abgeholt und von Prof. und Juden am Quai erwartet. Machen alle tropisch aufgeweichten Eindruck. Der Europäer braucht grösseren Stoffwechsel-Anreiz als diese ewig feucht-warme Atmosphäre bietet. Was hilft da Naturschönheit und Reichtum? Ich glaube, dass das Leben eines europäischen Arbeitssklaven immer noch reicher ist, vor allem weniger traumhaft und verschwommen. Anpassung wohl nur unter Preisgabe der Regsamkeit möglich.

5. Spaziergang in die Stadt mit Kohn, Typus Geschäftlhuber. Mittag mit dessen Frau

Also shows me a note about Parisian experiments on bacteriophages, which I definitely consider falsely interpreted.[92]

Picked up by hotel personnel in harbor and awaited by professors and Jews on the quay.[93] They all give the impression of being softened up by the tropics. The European needs a greater metabolic stimulus than this eternally muggy atmosphere offers. What good is natural beauty and wealth in this regard? I think that the life of a European slave laborer is still richer, above all, less dreamlike and hazy. Adaptation probably only possible at the price of alertness.

5. Stroll into the city with Kohn. A busybody type.[94] Noontime with his wife

24. Einstein with Jewish community president Isidoro Kohn, Rio de Janeiro, May 1925 (Courtesy Marcelo Gleiser, Dartmouth College, NH).

38

und deren Gesellschafterin (?) im Hotel. Die
Frauen nett und lustig. Nachmittags Besuch
u. Einladung einiger deutscher Kaufleute.
Darauf mit Professoren nach „Zuckerhut".
Schwindelnde Fahrt über wildem Wald
an Drahtseil. Oben prächtiges Wechselspiel
von Nebel und Sonne. Abends Begrüssung
von jüdischen Vereinen. Dann nächtliche
Autofahrt mit dem sehr sympathischen,
klugen und feinen Rabbiner Raffalovich.
6. Spaziergang mit Silvanello im
oberen Stadtteil. Kluger feiner Mensch,
der mich in die kleinen Intriguen der
Fakultät einführt. Die Sprache zieht
hier mehr als die Beobachtung. Mittag
in besserer Hafenkneipe. Scharfes Fisch-
gericht. Nachmittags Besuch bei
Präsident, Unterrichtsminister, Bürgermeister. Um 4 1/2 erster Vortrag im
Ing. Klub in überfülltem Saal

and her lady companion (?) at the hotel.[95] The women, nice and breezy. In the afternoon, visit & invitation from some German businessmen. Next, with professors to "Sugarloaf." Dizzying ride above wild forest by cable car. At the top, splendid interplay between fog and sun. In the evening, welcome by Jewish associations. Then nocturnal drive with the very likable, intelligent, and decent Rabbi Raffalovich.[96]

6. Stroll with Silvamello in upper part of the city. Clever, decent person, who introduced me to the little intrigues of the faculty. Language has greater pull here than observation. At midday, in superior harbor pub. Spicy fish dish.[97] In the afternoon, visiting the president, education minister, mayor. Around 4:30 p.m. first lecture in Engineers' Club in overcrowded hall

25. Einstein's first lecture on relativity at the Engineering Club, Rio de Janeiro, 6 May 1925, with Isidoro Kohn, journalist Assis Chateaubriand, and Admiral Gago Coutinho, *Careta*, 16 May 1925 (Courtesy Biblioteca Nacional, Rio de Janeiro).

39

bei Strassenlärm mit offenen Fenstern. Schon rein akustisch Verständigung unmöglich. Wenig wissenschaftlicher Sinn. Bin da so eine Art weisser Elefant für die andern, sie für mich Affen. Abends allein im Hotel auf meinem Zimmer gewisse nackt Blick auf die Bucht mit zahllosen grünen, teils nackten Felsinseln bei Mondschein.

7) Besuch im naturhist. Museum Tiere und Anthropol. Hauptsache Schönheit der Wirbelsäule einer Schlange als Konstruktion. Kultur der Indianer. verkleinerte Mumien, Giftpfeile. Herrlicher Garten vor Museum. Statistik der Rassenmischung. Schwarze verschwinden wegen mangelnder Resistenzkraft der Mulatten allmählich durch

with street noise through the open windows. For acoustic reasons alone, communication impossible. Little scientific sense.[98] Here I'm some kind of white elephant to the others, they are monkeys to me. In the evening alone in the hotel in my room naked I enjoy the view onto the bay with countless green, partly bare rock islands in the moonlight.

7) Visit to the natural history museum. Animals and anthropology the main focus. Beauty of the design of a snake's backbone. Culture of the Indians, miniaturized mummies, poisoned arrows. Magnificent garden in front of the museum. Statistics on racial mixture. Blacks gradually disappearing through mixing because of mulattoes lacking power of resistance.

26. Einstein at the National Museum of Brazil in front of the Bendegó meteorite, with engineer Alfredo Lisboa, president of the Brazilian Academy of Sciences, Henrique Morize, and mathematician Ignácio do Amaral, Rio de Janeiro, 7 May 1925 (Courtesy Biblioteca Nacional, Rio de Janeiro).

40

Mischung. Indianer relativ wenig
zahlreich. Mittags bei Prof. Castro.
Richtiger Affe, aber interessante Gesellschaft:
russischer Archäologe, kluger Journalist,
Schriftstellerin, hübsch, klug und etwas
arrogant. Nachmittags Akademie der
Wissenschaften. Kerle sind gewaltige Redner
Wenn sie einen rühmen, so rühmen
sie die – Beredsamkeit. Ich glaube,
dass solche Afferei und Unsachlichkeit
doch mit Klima zu thun hat. Die
Leute glauben es aber nicht.
8/ Morgens Bes. des Biologischen
Institutes. Path. Anatomie. Krankheits
übertragende Insekten. Trypanosomen
durch Mikroskop. Nachmittags
Vortr. in Ing. Schule. Grosse Hitze
in überfülltem Saal. Abends
Einladung in deutschen Klub
„Germania". Gemütliches Abendessen.

Indians relatively less numerous. At noon, at Prof. Castro's. Real ape, but interesting company: Russian archaeologist, smart journalist, female writer, pretty, intelligent, and somewhat arrogant. In the afternoon, Academy of Sciences. Those fellows are phenomenal speakers.[99] When they laud someone they laud—eloquence. I believe such tomfoolery and irrelevance really does have something to do with the climate. But the people don't think so.

8) Morning visit to the institute of biology. Pathological anatomy. Disease-transmitting insects. Trypanosoma through the microscope. In the afternoon, lecture at the School of Engineering. Great heat in overcrowded auditorium. In the evening, invitation to the German club "Germania." Cozy supper.[100]

27. Einstein at the Oswaldo Cruz Institute, with tropical expert Adolfo Lutz, director of the institute Carlos Chagas, engineer José Carneiro Felippe, and deputy director Leocádio Chaves, Rio de Janeiro, 8 May 1925 (Courtesy Archive of Casa de Oswaldo Cruz, Rio de Janeiro).

41

9) Sternwarte mit interessanten
Apparaten für Erdbeben.
~~Nachm.~~ Mittagessen bei Silvamello. Sehr behaglich mit
brasilianischen Speisen. Es
war sehr behaglich in diesem
Hause. Zu Fuss Besuch bei
zwei Brüdern, Physiologen
mit interessanten Arbeiten
über Athmung. Abendessen
bei Kohn. Ordinäre aber gutmütige
Menschen. Grosser Empfang der
Juden um 9 Uhr abend im Jokey-Klub.
Lange Reden mit viel Begeisterung
und unmässiger Lobhudelei,
aber alles redlich gemeint. Gottlob,
dass es vorbei ist. Noch zwei Tage
zu absolvieren, die dem Aussehen
nach angenehm sind. Aber unwider-
stehliche Sehnsucht nach

9) Observatory with interesting instruments for earthquakes. Lunch at Silvamello's. Very cozy, with Brazilian dishes. It was very comfortable in that house. Went by foot to visit two brothers, physiologists with interesting work on respiration. Dinner at Kohn's. Common but good-natured people. Major reception given by the Jews at 9 p.m. at the Jockey Club. Long speeches with much enthusiasm and inordinate adulation, but all meant sincerely. Praise be to God that it's over. Two days left to complete, which by appearances are pleasant ones. But irresistible yearning for

28. Einstein at the National Observatory, with astronomer Domingos Costa, Alfredo Lisboa, Alix Corrêa Lemos, director Henrique Morize, Isidoro Kohn, and Ignácio do Amaral, Rio de Janeiro, 9 May 1925 (Courtesy Fundo Observatório Nacional, Acervo do Arquivo de História da Ciência do Museu de Astronomia e Ciências Afins [MAST], Rio de Janeiro).

42

Ruhe von den vielen unbekannten Menschen. Abends nach Robert (auf Heimreise von Santos).

10. Herrlicher Ausflug mit Kohnstam. und Kommission per Auto auf verschiedene Aussichtspunkte. Sonnen-Untergang mit Zahnradbahn auf Corcovado. Abends Empfang in zionist. Centrale in gefülltem Saal bei ungeheurer Hitze. Ventilatoren nicht mehr fühlbar. Reden kürzer als gewöhnlich.

11. Besuch im Irrenhaus, dessen Direktor ein Mulatte und besonders vorzüglicher Mensch ist. Bei ihm brasilianisches Mittagessen mit viel Pfeffer und deutscher Frau. Dann Besuche bei Ministern, die gottlob meistenteils abwesend waren. Dann photographieren, Grosse Kino-Vorführung des Indianerlebens und deren vorbildliche Kultivierung

peace and quiet from so many unfamiliar people. Evening still with Robert (on his way home from Santos).[101]

10. Glorious excursion with Kohn's family and commission by car to various vista points. Sunset with cog railway up Corcovado. In the evening, reception held by the Zionist central office in a full hall in enormous heat.[102] Ventilation no longer palpable. Speeches shorter than usual.

11. Visit to insane asylum, whose director is a mulatto and an especially exceptional person. Brazilian lunch at his place with much pepper and German wife. Then visit to ministers, who, thank heavens, were mostly absent. Then photograph taking. Great film presentation of Indian life and their model cultivation

43

durch General Rondon, einen Menschenfreund und Führer ersten Ranges

Endlich Abendessen im Hotel gegeben vom deutschen Gesandten

Dann noch viele Briefe und Unterschriften. Endlich frei aber mehr tot als lebendig.

by General Rondon, a philanthropist and leader of the first order. Finally, dinner in the hotel, hosted by the German ambassador.[103] Then still many letters and signatures. Free, at last, but more dead than alive.

Additional Texts

TEXT 1. From Max Straus[1]

Berlin S. O. 33, Schlesischestr. 26, 5 November 1923

Esteemed Professor,

The proprietor of the Max Glücksmann firm in Buenos Aires, with whom my company has been cooperating for many years on a grand scale and who is a personal friend of mine, sent me a telegram some 10 days ago which reads as follows:

"For Straus, Glücksmann as representative of excellent Jewish families requests inviting Einstein visit Argentina; they extend guarantee of $4,000.–free passage there and back first class, sojourn free of charge, favorable time April, May, June, July, August, September; request reply by cable. Rosenbaum."

which I received only today, as I just now arrived back from a trip abroad.

I resp[ectfully] ask you please to inform me whether you can give me an opportunity to talk with you re. this telegram and, upon advance notice by telephone (Moritzplatz 11840), am ever pleased to place myself at your disposal.

I sign with my compliments in utmost respect,

M. Straus.

TEXT 2. From the Asociación Hebraica[2]

Buenos Aires, 9 January 1924

Dear Professor,

Mr. Straus, whom we have charged with transmitting to you our proposal of a voyage to the Republic of Argentina, has informed us of the conversation he has had on this subject with Mrs. Einstein, as well as the questions to which you requested clarification before making your decision.[3]

We hereby provide the clarifications requested by you.

1). The "Asociación Hebraica" is a society created with the sole aim of increasing the general level of intellectual culture among the Israelite community of Buenos Aires and, in particular, enhancing the study of Jewish history, literature, language, and philosophy.

All matters of a political or nationalistic nature are excluded from the program of the "Asociación Hebraica."[4]

2). According to your wish, which by the way entirely coincides with our own intentions, we have intervened with the University of Buenos Aires to the effect that they would invite you to deliver some scientific lectures in this country.[5]

In the attached extracts from Buenos Aires newspapers,[6] you will see that the university, together with the four other Argentine universities, has resolved to extend an invitation to you and has accepted to this end a donation from the "Asociación Hebraica" through which we intend to facilitate the financial issue.

You will see in these extracts that the University of Buenos Aires has decided to offer you, as honoraria, the sum of 4,000 dollars (including our donation) as well as two return passage tickets.[7]

The other Argentine universities on their part will certainly contribute as well.

The university has just informed us that their official invitation has already been sent to you through representatives of the Argentine Ministers in Germany and Holland.[8]

3). We have taken similar steps in the Republic of Uruguay, and the University of Montevideo has decided on its part to invite you, contributing the sum of 1,000 gold pesos.[9]

4). We believe that we can do the same in Chile, if you are interested in traveling to this country, which would enable you to make your return trip through the Pacific; in an affirmative case, please let us know without delay and we will take it upon ourselves to undertake the necessary steps to this effect with the University of Santiago.

5). We do not need to settle your schedule during your stay in Argentina and the other countries, since this will depend on your arrangements with the various universities.

On our part we have no obligations to impose upon you, since our sole aim, in taking the initiative of arranging your voyage, has been to fulfill one of the goals of our cultural intellectual program and at the same time to provide you the opportunity to visit South America.

Nevertheless, we would be very pleased to offer you a reception and to hear one or two popular lectures on a subject that you consider appropriate to the character of our institution.

But in no case should you fear any action on our part that would use your name for any kind of advertising or propaganda, since this would be contrary to the purely intellectual character of the "Asociación Hebraica."

6). Some Israelite notables are putting their homes at your disposal for the duration of your visit to Buenos Aires.

We therefore hope to be able to offer you the kind of comforts during your visit in our city that would allow you to work and rest in a satisfactory manner, unless the Argentine government, as is possible, will offer you an official welcome.

7). In case you are interested in the Israelite Argentine agricultural colony, the work of Baron de Hirsch, which has greatly flourished and is generally well known, the general director of the Argentine Colonies of the "Jewish Colonization Association," who is also the president of the "Asociación Hebraica," will be greatly pleased to invite you and arrange a visit of these colonies.[10]

We hope that our explanations will have satisfied you and that we will have the honor of personally conveying to you in Buenos Aires the respectful greetings of the "Asociación Hebraica" and to welcome you with the entire Argentine intelligentsia that has studied and admired your scientific works.

Please accept, Mr. Professor, our most devoted sentiments,

The Secretary — The President

M. Nirenstein[11] — [Isaac] Starkmeth

TEXT 3. To the Asociación Hebraica[12]

Berlin, 8 March 1924

I confirm herewith, with many thanks, the receipt of your letter of January 9th. This invitation delighted me so much that I most certainly feel like accepting it right away. Upon calm reflection, however, I had to say to myself that I can no longer travel to South America during the year 1924. First, I am currently engaged in scientific research here together with others and cannot bring myself to interrupt it. Second, in the last few years I have been absent from Berlin so much that I could not well justify to the local authorities traveling yet again for such a long time.[13]

In thanking you from my heart for your magnanimous invitation, I ask you to keep it open until it becomes possible for me to accept it. In return, I gladly promise not to accept any other invitation from abroad before I have visited you.

In utmost respect,

Albert Einstein.

TEXT 4. To Paul Ehrenfest[14]

[Berlin,] 12 July 1924

Dear Ehrenfest,

[. . .] I don't want to go to Pasadena in the foreseeable future.[15] The best refuge one can aspire to is "splendid isolation" from one's fellow human beings. But unfortunately, I'm going to have to travel to South America next June; they're practically skinning me alive. [. . .][16]

TEXT 5. To Maja Winteler-Einstein and Paul Winteler[17]

[Berlin, after 4 December 1924][18]

[. . .] At the beginning of March I'll be traveling to Argentina; they've been pestering me for years, and I have now relented (out of love for the sea). I hope to be back again at the end of May. [. . .]

TEXT 6. From Isaiah Raffalovich[19]

Rio de Janeiro, 27 January 1925

My dear Sir!

At the request of Dr. Aloysio de Castro, Director of the Faculty of Medicine in our University, (whom you met at the League of Nations Meeting at Geneva,) and Dr. Paulo de Frontin, Director of the Rio de Janeiro Polytechnic School and President of the Engineering Club, I had the honour of cabling you last week extending to you a cordial invitation on behalf of the above-mentioned Institutions to honour our Capital with your presence on your return from Argentine. The Jewish community, of which I have the honour to be the spiritual head, anticipate with pride and pleasure the honour which your presence will reflect upon them. We are looking forward to hearing that you have accepted the invitation and I pray you to honour me with your kind reply at your convenience in order to enable the Institutions to make the necessary arrangements for your reception.

I have the honour, My Dear Sir, to remain,

Yours Faithfully,
Dr. I. Raffalovich
Grande Rabbino.

TEXT 7. To Hermann Anschütz-Kaempfe[20]

[Berlin,] 17 February 1925

[. . .] On the 5th I'll be setting sail for Argentina. Long live the sea, but I'm not looking forward to the semi-cultured Indians there dressed in their tuxedos. I am returning at the end of May. [. . .]

TEXT 8. To Elsa Einstein[21]

Hotel [Hamburg, 5 March 1925]

Dear Else!

Trip appropriately magnificent. Blue sky and fine food, esp. figs. Mrs. Rob[inow] at train station with son-in-law.[22] Evening, small social circle. Melchior there too,[23] genuine Hamburger. Saw something of the city. Now it's time to board ship,[24] which casts off in an hour. Letter from Mühsam with cash.[25] Not yet opened.

Best regards esp. also to heroic Margot,[26] from your

Albert.

P. S. In my gray coat I forgot an envelope that contains a memorandum about Argentina.[27] Send it immediately out after me, because I need it in Buenos A[ires].

TEXT 9. To Elsa and Margot Einstein[28]

[on board "Cap Polonio,"] 7 March 1925

Dear loved ones!

Everything has proceeded very nicely up to now. After dispatch of my ticket from the hotel in Hamburg, I boarded the ship, accompanied by Mrs. Rob[inow]'s son-in-law,[29] where I was installed in my lordly bachelor's quarters, whereupon the voyage commenced immediately.

At first it was cold and it rained incessantly.—But I must report something else from Hamburg that was very funny. When I arrived at my hotel on Wednesday around 12 p.m., a small package was handed to me. In it was a black necktie and a visiting card: "ordered by telephone by Gov. Councillor Bärwald in Berlin." [30]—

Since yesterday it has been warmer; and now we've got fine weather. Yesterday we docked in the outer harbor of Boulogne where I dispatched a letter to Prof. Lewin;[31] I didn't have one to send to you yet. I often think how nice it would have been for Margot; but it was not meant to be. I'm sitting next to a German Professor of Philosophy who teaches at the university in Buenos Aires.[32] Otherwise I don't intend to make any acquaintances, and even this one, who is a very fine man, I merely see at mealtimes. I very much like the booklet that Rudi[33] had me take along. Besides that, I am busying myself leisurely with science. It turns out that I can't bear it otherwise, Lewin notwithstanding. When I try to leave it alone, life becomes too empty. No reading matter can substitute for it, not even scientific reading.

Now I'm outside a lot because the air along the French coast, that is, in the Bay of Biscay, is already very much milder than where we are. Definitely do immediately send me the exposition on Argentina from my gray coat, lower side pocket, that an Argentine (Gabiola) gave me in Berlin.[34] I'll need it in Buenos [Aires].

The ship is very large and rocks little. But strangely enough I feel it more strongly than on the earlier voyages. I'm going to leave the letter from Mühsam with the cash sealed until my return, provided I don't need any of the money; thus he must comply with my wish to leave me in peace. His conduct toward me proves that he knows me very poorly.[35]

The food on the ship is very hygienic, and I am extremely careful, because I am very well aware of the riskiness of my mission.[36] As beautiful as the sea voyage is, I do still consider it foolish of me to have let myself be drawn into this undertaking, with my predisposition. For over there I have the choice between much pestering and agitation as a consequence of annoyance and disappointment. I'm not spending much time perusing the Spanish booklet; my stomach rebels against that twaddle in its fourth edition, which I am already familiar with in three editions.

I just devoured pineapple in honor of poor little Margot. What a shame that she couldn't order it herself. I hope she is feeling better again.

Warm regards to both of you, the grandparents, and Rudilse[37] from your

Albert.

TEXT 10. To Elsa Einstein[38]

Cap Polonio, 15 March 1925

Einstein, Haberlandstrasse, Berlin

Near equator healthy and sunburned thanks for birthday letter[39]
Greetings to all

Albert

TEXT 11. To Elsa and Margot Einstein[40]

[Cap Polonio] 20 March [1925]

Dear loved ones,

Now the happy-go-lucky life is over. Because tomorrow we'll be reaching Rio at an ungodly hour, where God only knows what's awaiting me. The voyage was wonderfully relaxing. I had no more human interactions than was pleasant. A couple of nice musicians with whom I played, and a couple of people with whom I could converse interestingly. There is a woman writer, Mrs. Jerusalem, a professor,[41] and—a Bavarian priest. Add to that the captain, who is an uncommonly witty eccentric. Otherwise I paid little tribute to society. I attended dinner as a single male, appearing just as I was—and that was not always palatable—and avoided people. In return, I paid tribute by sawing away on the fiddle for the people at a festive concert, which evoked great glee.[42]

The heat is far lower than in the Red Sea;[43] and I was always very abstemious and cautious with the meals, so not the slightest thing happened to me. It sometimes does demand willpower because the cooking is excellent. I didn't work a lot due to the heat, as opposed to the Japan trip. I feel very equal to the exertions ahead of me and view what's ahead with serenity, although without much interest, either. That little Margot[44] could not come along I regret with each experience that would have been nice for her. To be sure, there will be many more like it. Sunshine, stars, sea, flying fish, dolphin, a charming tour through filthy, dreamy Lisbon, the passage along the mighty Pic a Tenerife[45] (similar to Mt. Fuji) and much else.

I hope all of you are well and everything else is in order, too. Don't have the salary for May picked up, d[ear] Else. I'll arrange the matter after my return. In the meantime, telephone Planck[46] accordingly, so that he knows.

Warm regards all around, also to the grandparents, from your

Albert.

TEXT 12. To Elsa and Margot Einstein[47]

Buenos Aires [26 March 1925][48]

Dear loved ones,

Yesterday, after the steamship got stuck, arrival, in the best of health, in Buenos Aires with a great hullabaloo. I am living at Mr. Wassermann's home (Sabala 1848).[49] By the time you get this little note, I'll already be in Montevideo or Rio, from where, on May 12, I'll be traveling back toward Hamburg. I am lodging with a very likable family and am protected against all intrusions. The schedule is immensely packed, but I feel strong and indifferent toward people. Because what

I'm doing here is probably little more than a comedy. Buenos [Aires] is a barren city from the point of view of romanticism and intellectuality. But I am enchanted with Rio.

Some fellows are waiting for me again. I hope you are as healthy as I am and have smaller paunches. Heartfelt greetings to you, the grandparents,[50] and friends from your

Albert.

I'll probably hardly be able to write any picture postcards despite the fine addresses I have here. I briefly saw Dr. Silvamello in Rio, tell Ehrmann I did.[51] On my return to Rio, where I'll be staying for 6 days, I shall see more of him. Also greet Katzenstein and my dear old Lewin.[52]

TEXT 13. Statement on Zionism[53]

[Buenos Aires, 28 March 1925]

[. . .] I am not a Zionist in the sense that I believe it will resolve the Jewish problem. I sympathize with the work done by the supporters of this ideal and I participate in this work because I am convinced of the necessity of creating a moral center that could unify all Jews and, where feasible, disseminate Jewish culture. The latter is the role of the University of Jerusalem which will be inaugurated shortly.[54] [. . .]

TEXT 14. To Elsa and Margot Einstein[55]

[Buenos Aires,] 3 April 1925

Dear loved ones,

Yesterday I received your first letter, dear Else, with the many reports of illness. I hope everyone is well again, as it should be. On the ship I regretted very much not having taken Margot along, but now it has proved to be the right decision, nevertheless. I think I'll convert all of you against medicine, after all! I now have one week of razzmatazz in Buenos Aires behind me. The finest thing was that on 1 Apr. I flew 1,000 m high above Buenos Aires in an airplane.[56] People are very nice and kind to me and the Wassermann family makes life easier for me in every way.[57] If I were living alone and unprotected in a hotel, I could not have endured it. One week (from the 7th to 14th[)], I shall have my peace almost entirely, apart from a trip to Cordoba on 11 Apr.[58] On 23 Apr. I'll be traveling to Montevideo and around 1 May to Rio, whence my ship departs back to Hamburg on the 12th. If only it were that far already. This farce is actually wholly uninteresting and quite strenuous. I've already spent one evening visiting Robert and am going to be there on another evening again (Saturday); his boss L. Dreyfus is here.[59] He owns a billion francs and is so cheap as to sleep in Robert's bed instead of in a hotel. I will not embark on another such trip again, not even if it's compensated better; it's one big drudgery. The country here is, oddly, exactly as I had imagined: New York, mellowed somewhat by southern European races, but precisely as superficial and soulless. I don't want to be here. I'll be going to Cordoba for just a short while, so I'll scarcely have anything to do with N[icolai].[60] Your Mrs. Hirsch[61] didn't know who you were, but finally I did manage to make contact. It's a house full of precious artwork, dripping in wealth. Her husband is one of the most powerful moneybags in Argentina,[62] an interesting person to look at—once. The Argentine minister of foreign affairs[63] comes regularly to my lectures—as a sign of genuine democracy, where nobody is too good for learning. But on the whole, nothing but money and power counts here, as in North America. But there are better ones among the young, as in America.

Warm regards from your

Albert.

I have half the lectures behind me, thank God. I hope all of you are in good health.

TEXT 15. On the Inauguration of the Hebrew University of Jerusalem[64]

[Buenos Aires, 6 April 1925]

Jewish Community Celebrates the Founding of the University of Jerusalem

In celebration of the recent founding of the first Jewish University in Palestine, a formal tribute was organized by the Argentine Zionist Federation last night at the Teatro Coliseo, whose ample auditorium was completely filled. [. . .][65]

Professor Einstein's Lecture

The renowned sage delivered a brief lecture in German, dealing primarily with the Zionist movement.

He said that Zionism has restored the dignity of the Jews, which previously had been perceived as diminished and denigrated. He then declared himself a supporter of the founder of Zionism, Dr. Theodor Herzel, and the current leader, Dr. Weitzman, who, he said, is an inspired politician.[66]

He added that the Jewish people are not fighting to rebuild their nationality in an aggressive nationalistic way, but rather with the noble, human purpose of making their special culture arise again as part of universal culture.

He stated that the recently created University in Jerusalem is one of the fundamental elements of the cultural enterprise sponsored by Zionism, which should be supported by everyone, since enterprises of such magnitude need the stimulus of collective effort.

He remarked that Judaism in Western Europe is dissolved. Some great personalities have emerged there, he added, but as a nucleus it lacks spiritual meaning.

We must prevent this process from continuing, and that will be achieved only through the establishment of Zionism, which is the only possible salvation of the Jewish soul.

It is our mission to struggle with all means at our disposal so that Zionism may prosper, and in this way we will save our culture.

Professor Einstein was cheered several times with enthusiastic applause.

TEXT 16. To Ilse Kayser-Einstein and Rudolf Kayser[67]

[Buenos Aires,] 8 April [1925]

Dear loved ones,

After a glorious voyage and much preaching and acquainting myself with Argentines, Jews, and physicists, I am sitting here for a couple of days taking a breather on an isolated country estate.[68] Then it's onwards to Cordoba and later (Apr. 22) to Montevideo, then May 2 to Rio, and on May 12 on board the return ship; on May 31 I'm arriving in Hamburg, if all goes smoothly. Give my greetings also to the Haberlanders;[69] I don't have enough postcards, otherwise I would have written to everyone.

Warm regards, yours,

Albert.

TEXT 17. To Margot Einstein[70]

[Buenos Aires,] 10 April 1925

Dear little Margot,

I hope you're doing fine. I spent three beautiful days on Wassermann's country estate[71]—Isle of the Blessed. But then it's going to be completely frantically busy again (Cordoba, Montevideo, Rio), a real test of the nerves. On May 31 I'm arriving back in Hamburg. I've been managing quite well up to now and mangling French to blazes like the devil, so that sparks have been flying.

Fond greetings to all of you from your

Albert.

TEXT 18. To Elsa and Margot Einstein[72]

[Buenos Aires,] 15 April 1925

Dear loved ones,

What a lot of things I've experienced! You'll read about it in my journal.[73] All in all, it went quite well, but my head is as if stirred up inside with a ladle. If the Wassermanns[74] hadn't protected me so well, I would surely have gone nuts; this way, only halfway so. In one week I'll be traveling to Montevideo, then around the 11th of May to Rio.[75] On the 11th my ship departs from there for Hamburg.[76] Health-wise I've survived it all well. People have been very good to me all round. The German colony ignores me completely, which is the simplest for me; they seem to be even more nationalistic and anti-Semitic than in Germany proper.[77] The German envoy,[78] however, was very solicitous toward me; he is being boycotted by the Germans here because he's liberal. One generally laughs about the political foolishness of the Germans here.

I'm glad that all of you are properly on your feet again. I received your news from 18 March yesterday upon my return from Cordoba. I was very decent towards Nicolai. He did not ask about his little son, and so I didn't say anything about him either.[79] He still wears a monocle, has aged considerably, still has a big mouth and—fits in very well in Cordoba. But the more refined minds don't take him seriously here, either. Up to now I have eight lectures and oodles of official dinners behind me. One time it affected my intestines, in La Plata after having eaten at the "Jockey Club."[80] Thereupon I said to the wise men of Buenos Aires that I must have eaten some part of a race horse, because I had to run so much afterwards. This fine joke earned me much sympathy because jokes are very much loved over here. The newspapers are as impertinent and intrusive as in North America. Overall there are, despite the racial differences among the inhabitants, great similarities, which are explained by the intermingling of the population, the natural wealth of the country. But there are also more down-to-earth things, a kind of folk music that interested me very much. A horrendous amount of meat is devoured here, which is generally of very good quality, and has lent my figure an undesirable ampleness. I do apply the brakes as much as possible, though. I still have to endure almost four weeks; I dread it a little. I was on Wassermann's country estate in Llavajol for three days in complete tranquility, and had an exceedingly valuable scientific idea there that, peculiarly enough, draws on what I found during the return trip from Japan.[81] All in all, I must admit that traveling without a wife under such circumstances is simpler because there is less socializing fuss.

Now it's only about six weeks until I come home. I've already sent off quite a few picture postcards but not enough yet. I'm hoping to be able to manage with that, too. I'm curious what Ehrmann has to say about Palestine.[82] You've probably already heard about the large donation for a physics institute in Jer[usalem].[83]

Heartfelt greetings from your

Albert.

Don't part with the manuscript, dear Else.[84] Ehrmann is entirely right. It's not a good time for the sale. Better directly after my death. I've received a very fine payment here,[85] so from that point of view it isn't all for naught.

TEXT 19. Statement on Nationalism and Zionism[86]

[Buenos Aires, 18 April 1925]

[. . .] According to my ideological point of view, an undivided humanity would be desirable; however this seems impossible today, nor can one envision it for tomorrow; nationalism is justified as a practical fact and thus Zionism is also justified. Zionism is a dignified solution for a moral, cultural and national state, not just for those Jews embracing Palestine as their nation, but also for Judaism in general; it is a motive for cohesion and a reason for being. [. . .]

TEXT 20. To Elsa and Margot Einstein[87]

Buenos Aires, 23 April 1925

Dear loved ones!

Now the program in Buenos Aires is exhausted, but so am I. This evening it's onwards to Montevideo. Without the Wassermanns[88] I couldn't have made it but would have been devoured instead by sheer love. I'm earning over 20 M. in Montevideo and Rio, hence, very decently.[89] But I'll never do such a thing again. It's such a terrible slog. But I have stayed completely healthy in the process, I just got a bit—fat. I received lovely letters from you and Rudilse and was delighted, most particularly with Margot's.[90] I'm making this letter brief; and you'll have to console yourselves with the diary, because I don't have any time to narrate. Our Jews are importuning me the most with their

29. Members of the Sephardi Jewish community bidding farewell to Einstein, Buenos Aires, 23 April 1925 (Courtesy Archivo General de la Nación, Buenos Aires).

love. I was able to be very decently effective for the Zionists.[91] The cause is powerfully gaining ground here, too. Just now I'm back from a small reception by the Sephardi Jews in their temple, which was so beautiful that I had to cry.[92] Hardly a word was spoken throughout—strange.

I'm being called away. More another time. But this letter must go out, because I don't know when I'll be getting around to writing again. I'm glad to be rid of the manuscript now and thank you for the labor of love; better this way than burned or sold.[93] Especially the latter would have been undignified.

Best regards to the grandparents, [94] to both of you, and to Rudilse from your

Albert.

TEXT 21. To Hans Albert and Eduard Einstein[95]

Buenos Aires, 23 April 1925

Dear Boys!

I've been in Buenos Aires for almost one month now and have had an enormously strenuous existence; I was also in Cordoba (have a look at it on the map!) Now it's still onwards to Montevideo and Rio. End of May I'll be returning home. I would so much have liked to take one of you along with me, had school allowed. But rarely can one do as one pleases; one is always *obligated*. I'm sending along a couple of pretty postage stamps and will see that I get a few cacti in Rio for Mama.[96] As for me personally, I found an interesting idea for understanding the connection between electricity and gravitation.[97] I'm already looking forward to our being together during the summer. In addition to visiting Kiel,[98] I also have to spend some time at a high altitude because the doctor absolutely demands it; I don't yet know when that will be, though.[99] Write to me to Berlin. I'd be so happy to find a letter there from both of you.

Warm regards from your

Papa.

30. Einstein visiting the Montevideo College of Engineering, with students, 29 April 1925 (Courtesy Leo Baeck Institute, New York).

TEXT 22. To Elsa and Margot Einstein[100]

Montevideo, 27 April 1925

Dear loved ones!

Now I've been here in Montevideo, a beautifully situated harbor city, for half a week already. It is much cozier here than in Buenos A[ires]. The city is smaller, more handsomely laid out. I am living with a Russian Jew who can only speak Yiddish and Spanish.[101] I'm being overwhelmed with so many honors, that I can hardly catch my breath anymore, but in such an ingenuous way. This is supposed to continue for the whole week. The ship to Rio[102] leaves on Saturday May 2, and on the 12th from there to Hamburg. Today I have to visit the foreign minister and the president before the lecture, whereas this morning the president of the Senate gallivanted around with me for two hours.[103] Over here the German colony is behaving more politely, after the one in B[uenos] A[ires] seriously disgraced itself by its decision to ignore me.[104] But that just means one more burden for me. Yesterday the mayor drove me around and showed me the hotels that the city itself has built.[105] There they're running roulette à la Monte Carlo with much success. This little country is dripping with riches. It has suddenly become so cold that one is reminded of Europe. It's going to be all the hotter in Rio, which has a completely tropical climate. I've been pampered here as virtually never before in my life, but my brain feels as if it's been stirred with a ladle, so it seems hard to think of it ever being of use for anything reasonable again. But I *do* still hope so. It perhaps is better, after all, that you aren't here with me, dear Margot,[106] because such a long time full of social engagements would have gotten on your nerves a lot, even more than for me.

Fond greetings to you all, as nephew, son-in-law, husband, stepfather and father-in-law.

Yours,

Albert.

I'll be getting your letters only in Rio on the ship. I'm still healthy and cheerful.

TEXT 23. "On Ideals"[107]

[Buenos Aires, 28 April 1925]

All lives, both individual and collective, after satisfying their most common material needs, yearn for a world of superior values which, because of its reactive effect on men, tends to make them nobler and more spiritual. All things begin with myth, primitive religion, pure animism, the deification of nature and the forces that control it. But in the subsequent development of the "European" people, the determination of life's values and ideals is in no sense limited to religion; rather, through constant growth, as well as historical evolution, it becomes

31. Einstein at writing desk with bust of Dante, Buenos Aires, late March–April 1925 (Courtesy Leo Baeck Institute, New York).

externalized, most notably in the literary, artistic, and philosophical life of nations. This fact, or rather, these fields of intellectual production, which for the most part are essentially subjective and perhaps the principal ideals of life, differ markedly from Eastern thought,[108] in which the great religious systems and compendiums of knowledge, of equally religious character, constitute almost all of superior and superhuman truth, and with their pretensions of eternal validity and objectivity, exclude historical evolution as well as a multiplicity of doctrines and those who teach them. Linked to this formal distinction is also the essential one: the European ideal of life tends, first and foremost, to produce a "great and unique personality," set apart from the crowd and from the present moment. The quintessential European ideal is that of "the hero, the fighter," and its devotion to the world of ideals beyond the material is practically the equivalent of a "veneration of heroes" tinged with religious overtones. This explains the mythical character acquired by men like Caesar and Napoleon, but spiritual creators—a Dante, a Goethe, a Nietzsche—can also take on heroic proportions in the conscience of the people. The Asian ideal, unlike the European, entirely disregards the man of action and his culmination in the heroic. In this regard Taoism, Buddhism, Judaism, and Christianity completely coincide. For them, doctrine comes first, devotion to an idea that is valid for all, the recommendation of a morally pure life. The contradiction resulting from this contrast between East and West can be noted throughout the history of European Christianity, so that despite the widespread Christianization of the continent, the Christian ideal of life never managed to prevail to the exclusion of all others. This ideal has a passive character and is at odds with active Europeanism, which, when taken to an extreme, tends to create great personalities, heroism, and individual productivity.

This Europeanism is already clearly evident in Hellenism. Devotion to the ideals of beauty and truth is the manifestation of an active creative spirit, at the same time requiring the establishment of values that respond to such a psychology. Plato's world of ideas, with its closeness to the religious East, is absolutely European insofar as the establishment of different values is concerned. No matter how much the ideals and values of life may have changed in the history of Europe, they still retain this active, productive character. Moreover, European heroism is apparent even in those spiritual currents that oppose it by nature (here I see, for example, the difference in principle between Asian and European mysticism).

The newest "American" ideal of life is as different from the Asian as it is from the European. It does not tend toward the creation of meta-psychic moral values; neither is it hero worship or faith in personalities. Rather, its objective is economic power. The progressive

Americanization of Europe gives this ideal a pronounced practical validity on our continent as well, even though it may be denied and contested as an ideal value, because it is the opposite of the European ideal. For an American, the only criterion by which all values are judged is practical reality, a concept of life that had its philosophical expression in pragmatism, whose underlying idea goes like this: "The truth is whatever can be proved in praxis."[109]

Despite this contrast, as far as spiritual and artistic life is concerned, I do not see any antagonism between Europe and America. The psyche of both worlds tends toward productive growth. In America a high level of technical and economic acumen prevails; however, I do not believe that this excludes all spiritual life. Technical and economic action, after all, also provides room for creative gifts, since genius can overcome mechanical rules and develop them freely. Furthermore, the most rigorous organization of economic life creates possibilities for freeing spiritual creators from material concerns.

For the Western world (Europe and America), the religious ideal of the East calls for productivity. He who truly adopts this ideal in the spiritual, artistic realm of life, does not, in my opinion, need any other ideal whatsoever and cannot therefore propose any objective for himself other than productive development. "He who possesses science and art also possesses religion; he who does not possess them needs religion" (Goethe).[110]

TEXT 24. To Carlos Vaz Ferreira[111]

[Montevideo,] 29 April [1925]

Dear Mr. Vaz Ferreira,

Thank you very much for the precious gift you gave me.[112] I have already begun to read your work on pragmatism.[113] I am not a pragmatist. I find that it gives quite an imperfect definition of the truth.

32. Einstein and Uruguayan writer Carlos Vaz Ferreira, Montevideo, 24 April 1925 (Courtesy Leo Baeck Institute, New York).

But were I a pragmatist, I would respond to your criticism of pragmatism[114] in the following manner: "I do not give a definition of *the* truth because *the* truth doesn't exist. One can only give a definition of 'truth of a statement in relation to a given and well-determined (limited) complex of consequences.' A statement that is 'true' relative to a certain system restrained by consequences is no longer true relative to a system further expanded by consequences."

I add that I do not see the problem in this way; but if one considers things in this manner, then the error that you mention disappears. But I concede to you that this remark changes nothing for your criticism of the use that James makes of his doctrine.[115] If one is careful of the truth, if it is *practical* to draw from a statement all possible consequences and thereby check the "truths," pragmatism offers us no new means by which to choose or to judge. Furthermore, one observes that the definition of the pragmatism of truth is insufficient, because it does not define the meaning of the word "consequences" (which ought to be "practical"). If one tries to follow up from this point of view, then one easily perceives that the principal difficulty that is encountered, if one wants to define the truth, is not resolved—not even touched—by the pragmatist theory. I am very sorry that I no longer have the possibility to address all these questions with you in person because of numerous social demands. I greet you with all my heart.

Yours,

A. Einstein

P. S. Excuse the horrible French. But I believe that you prefer it to a letter in German.
P. S. I think that the concept of "truth" cannot be addressed separately from the problem of "reality."[116]

TEXT 25. To Paul Ehrenfest[117]

[Rio de Janeiro,] 5 May 1925[118]

Dear Ehrenfests!

I've been roaming around this hemisphere as a traveler in relativity for two months already.[119] Here it's a true paradise and a cheerful mixture of little folks. On the 12th I'm homeward bound again.

Warm regards to all from your

Einstein.

Is Tania back already?[120] How did she like it?

TEXT 26. To Elsa and Margot Einstein[121]

[Rio de Janeiro,] 5 May 1925

Arrived here yesterday on French vessel in splendid weather.[122] Indescribable magnificence of the scenery. Staying at hotel.[123] Received two letters from you, dear Else. Am arriving on 31 May in Hamburg on board *Cap Norte*; departure here 12 May. Am glad that all of you are finally healthy again. What's wrong with Ilse?[124] What does Levin have to say about her case?[125] I'd like to know what he's got against F. H. objectively.[126]

Fond regards to all of you, your,

Albert.

TEXT 27. Radio Address for Rádio Sociedade[127]

[Rio de Janeiro, 7 May 1925]

My visit to Rádio Sociedade has led me, once again, to admire the splendid results achieved by science, in alliance with technology, and the transmission of the best fruits of civilization to those who live in isolation. It is true that a book could achieve the same, and, indeed, does, but not with the simplicity and assurance of a broadcast, carefully presented and heard live. A book must be selected which can sometimes lead to difficulties. When culture is shared through radiotelephony and individuals are sufficiently competent to broadcast, those listening receive a judicious selection, personal opinions, and commentaries that facilitate comprehension. This is the great work of the Rádio Sociedade.

TEXT 28. To the Jewish Community in Rio de Janeiro[128]

Rio de Janeiro, 11 May 1925

I feel the need to express once again my sense of gratitude to the local Jews on the occasion of my being named as honorary president of the Jewish Community and member of the Federation Sionista and the Jewish Library of Rio de Janeiro.[129] I would like to take this

33. Einstein arriving at the Automobile Club for the Jewish community reception, with Isidoro Kohn, Rio de Janeiro, 9 May 1925. *O Malho*, 16 May 1925 (Courtesy Biblioteca Nacional, Rio de Janeiro).

opportunity to encourage all the Jews here to support with all their might the endeavors of the admirable and tireless Rabbi Raffalovich.[130]

With deep respect,

A. Einstein.

TEXT 29. To the Chairman of the Norwegian Nobel Committee[131]

[On board "Cap Norte,"] 22 May 1925

Highly Esteemed Sir,[132]

I would like to take the liberty of directing your attention to the activities of General Rondon,[133] Rio de Janeiro, because during my visit in Brazil I gained the impression that this man would be highly worthy of the award of the Nobel Peace Prize. His work consists of the incorporation of Indian tribes into civilized humanity without the use of either weapons or coercion of any kind. My information derives from professors of the Polytechnic University in Rio de Janeiro, who spoke about this man and his work most warmly. I was also shown some things recorded on film.[134] I have not made the personal acquaintance of General Rondon.

I can provide you with more specifics upon request, but would prefer to see that you procure such information directly—perhaps through your Norwegian envoy.

With utmost respect,

Prof. Dr. A. Einstein

TEXT 30. To Mileva Einstein-Marić[135]

Bilbao, 27 May 1925

Dear Mileva,

The South American lecture tour is now behind me, thank God, and I'm returning home again on the 21st.[136] From Rio I brought back a little basketful of cacti that the director of the botanical garden[137] gave me to take along for you. They are planted in little pots. I'm taking them to Berlin for the time being, and the only difficulty is how I should get them into your hands. By the end of July at the latest there will surely be a session of the League of Nations committee.[138] If it takes place in Geneva, I'll bring the plants along to all of you myself and stop by in Zurich. Then I can also test what it's like to sleep in your house. Then I'll also bring along a collection of Brazilian butterflies for the boys[139] that I likewise received in Rio. I hope I bring them home intact; I will make every effort to do so. [. . .]

TEXT 31. To Karl Glitscher[140]

Bilbao, 27 May 1925

Dear Dr. Glitscher!

I was on a lecture tour in South America and am now on the return journey. Over there, one finds more well-tailored clothes than fine and interesting fellows who wear them. It still is better and more interesting at home in Europe—despite the various European follies, political and otherwise. [. . .]

TEXT 32. To Michele Besso[141]

Berlin, 5 June 1925

Dear Michele!

On 1 June, I returned from South America. It was a lot of agitation without much of any actual interest but at least a few weeks of peace and quiet during the sea voyage. [. . .]

To find Europe enjoyable, one must visit America. To be sure, the people there are freer from prejudices, but, at the same time, mostly mindless and uninteresting, more so than here. Wherever I go, I am enthusiastically welcomed by the Jews as, I am, for them, a sort of symbol of the cooperation of Jews. I thoroughly enjoy this, as I anticipate much pleasure from the unification of the Jews.

Warm regards to all of you from your

TEXT 33. To Mileva Einstein-Marić, Hans Albert, and Eduard Einstein[142]

[Berlin,] 13 June 1925

Dear Mileva and dear boys!

Thank you for all the enjoyable and lovely letters. The cacti were difficult to transport, but apparently arrived in good condition and are now under Margot's care in pots on the balcony.[143] The concern now is how we get them to Zurich. [. . .]

I think that Tete will rest as well in Kiel as in the mountains, especially considering that we will spend a lot of time on the sailboat.[144] I myself am not much in need of recovery despite the terrible exertions in South America, as the return voyage was already a recovery. Whether the butterfly collection withstood the trip well, I do not know, since I haven't taken it out of its meticulous packaging so that it is ready for immediate transport to Zurich.[145] Who knows whether it looks like a bit of potato flour? I hope not. [. . .]

Warm regards to all of you from your

Papa.

TEXT 34. To Maja Winteler-Einstein[146]

Berlin, 12 July 1925

Dear Sister,

[. . .] South America was a mad rush, almost unmanageable. I won't do that sort of thing again, too detrimental for the nerves. But the ocean voyage was splendid, and the Brazilian coast, with its fairytale-like forest and the conglomeration of peoples and the blazing sun. Now I have here at home a kind of small ethnographic museum, the most beautiful and loveliest things from Japan. They are a people with a deep, tender soul, in contrast to Argentina, which seems so banal and vulgar.[147] Everywhere I promoted the Zionist cause, and was received by the Jews with indescribable warmth. In contrast, the Germans boycotted me in Buenos Aires.

TEXT 35. To Robert A. Millikan[148]

[Berlin,] 13 July 1925

Dear Prof. Millikan!

It leaves a sour taste in my mouth to have to write this letter. For I must unfortunately inform you that I probably will be unable to come to Pasadena at all.[149] I was, in fact, in South America this year, and this trip had such a bad impact on my nerves that the doctor very strongly advises me against engaging in such a major undertaking in the next few years. I am very sorry to have to tell you this, since my respect for the scientific work you are doing there is very great, and my personal affection for you and the rest of the colleagues there is no less great.—[. . .]

With the hope of seeing you again soon, I am, with warm regards, your

A. Einstein

Chronology of Trip

1925

March 4	Arrives in Hamburg.
March 5	Departs for South America from Hamburg on board the S.S. *Cap Polonio*.
March 6	Docks at Boulogne-sur-Mer, France.
March 8	Docks at Bilbao, Spain.
March 9	Passes through A Coruña and Vigo, Spain.
March 11	Visits Lisbon.
March 12	Passes Tenerife.
March 14	Passes Fogo, Cap Verde.
March 21	Arrival in Rio de Janeiro. Tour of city by car.
	Visits Botanical Garden.
	Lunch at the Copacabana Palace Hotel, hosted by newspaper owner Assis Chateaubriand.
March 24	Arrival in Montevideo.
March 25	Arrival in Buenos Aires. Lodges at the residence of the paper merchant Bruno Wassermann.
	Visited by Leopoldo Lugones.
	Meets with German ambassador Carl Gneist.

	Visits rector José Arce and dean Eduardo Huergo of the University of Buenos Aires.
March 26	Tours Buenos Aires, including parks and fruit and vegetable market.
	Meets with the dean of the University of La Plata, Julio R. Castiñeiras, and with representatives of the Jewish community.
March 27	Reception and introductory lecture at the Colegio Nacional de Buenos Aires high school.
	Evening reception at residence of Alfredo Hirsch.
March 28	First scientific lecture at the University of Buenos Aires.
	Statement on Zionism is published (Text 13 in the Additional Texts section).
March 30	Second lecture at the University of Buenos Aires.
	Visits the offices of *La Prensa*.
	Meets with Robert Koch.
March 31	Visits the editorial office of *Dos Volk* and the Jewish quarter.
	Visits country estate of Hofer at "El Tigre."
April 1	Flight over Buenos Aires on a German hydroplane together with Berta Wassermann-Bornberg.
	Afternoon visit at the Ministry of Foreign Affairs, then meets with the president of Argentina, Marcelo T. de Alvear.

	Visit at the Ethnological Museum.
	Third lecture at the University of Buenos Aires.
	Dinner with Leopoldo Lugones.
April 2	Visits the University of La Plata and the Museum of Natural History.
	Luncheon at the Jockey Club, hosted by university president Benito N. Anchorena.
	Scientific session held in Einstein's honor at the university.
April 3	Luncheon at the Jockey Club in Buenos Aires, hosted by rector José Arce.
	Fourth lecture at the University of Buenos Aires.
April 4	Lecture on the occasion of the inauguration of classes at the Faculty of Philosophy and Letters at the University of Buenos Aires.
	Meets with Robert Koch and Louis Louis-Dreyfus.
April 5	Trip to Wassermann's *estancia* in Llavallol.
April 6	Visits ophthalmologist Eugenio Pablo Fortin with physiologist Bernardo A. Houssay.
	Fifth lecture at the University of Buenos Aires.
	Celebration of the inauguration of the Hebrew University in Jerusalem by the Jewish community at the Teatro Coliseo. Delivers speech (Text 15 in the Additional Texts section).

April 7	Visit at the University Clinic of Anatomy and Surgery, accompanied by its director, rector José Arce.
April 8–10	Spends Easter break at the Wassermanns' *estancia* at Llavallol.
April 11	Travels to Córdoba. Lodges at the Plaza Hotel.
between April 11–14	Meets with Georg F. Nicolai.
April 12	Travels to the Sierras de Córdoba mountain range.
	Banquet held in Einstein's honor at the Plaza Hotel, hosted by rector Léon S. Morra.
April 13	Festive session, lecture, and luncheon at the University of Córdoba.
April 14	Returns to Buenos Aires.
April 15	Sixth lecture at the University of Buenos Aires.
April 16	Attends session of the executive committee of the Federación Sionista Argentina.
	Makes Statement on Nationalism and Zionism (Text 19 in the Additional Texts section).
	Afternoon special session of the National Academy of Exact, Physical, and Natural Sciences to award Einstein honorary membership.
April 17	Seventh lecture at the University of Buenos Aires.
	Evening reception at the German Embassy.
April 18	Private lecture at the Wassermanns' residence.

	Reception in Einstein's honor hosted by the Jewish community at the Teatro Capitol. Lecture on the situation of the Jews.
	Reception at headquarters of the Asociación Hebraica. Awarded honorary membership.
April 19	Excursion to Llavallol.
	Visits Jacobo Saslavsky with Robert Koch.
	Dinner at the Hotel Savoy hosted by Zionist dignitaries.
April 20	Eighth (and final) lecture at the University of Buenos Aires.
April 21	Reception at Jewish hospital.
	Luncheon at the Rowing Club Tigre, hosted by dean Eduardo Huergo.
April 22	Official breakfast with academics and politicians including Foreign Minister Ángel Gallardo, rector José Arce, and the German ambassador.
	Banquet in Einstein's honor held by the Centro de Estudiantes de Ingeniería at the YMCA.
April 23	Lunch with Leopoldo Lugones and physicists.
	Visits Sephardi synagogue.
	Interview with *La Prensa*.
	Departure from Buenos Aires on board the S.S. *Ciudad de Buenos Aires*.

April 24	Arrival in Montevideo. Lodges at residence of Naum Rossenblatt.
	Visited by German ambassador, Arthur Schmidt-Elskop.
	Meets philosopher Carlos Vaz Ferreira on stroll.
April 25	Meets with delegation from local Jewish community.
	Tour of city by car.
	First lecture at the University of the Republic. Reception hosted by rector Elías Regules.
	Attends a performance of *La Traviata*.
April 26	Paul von Hindenburg elected president of Germany.
	Press conference at Rossenblatt residence.
	Stroll on coast with the president of the administrative council of Montevideo, Luis P. Ponce.
	Attends a performance of *Lohengrin*.
April 27	Tours the Compañia de Materiales para Construcción and the almost completed Legislative Palace with president of the National Senate, Juan Antonio Buero.
	Audience with president of the Republic, José Serrato.
	Meetings with minister of justice and public instruction and with Swiss consul.
	Second lecture at the University of the Republic.

April 28	Reception by Federation of German Associations at the German Club.
	Reception in Einstein's honor hosted by Jewish community at the Hotel del Prado.
	"On Ideals" (Text 23 in Additional Texts section) is published.
April 29	Reception and tour at the College of Engineering.
	Visits National Senate.
	Third (and final) lecture at the University of the Republic.
	Reception at residence of the German ambassador.
April 30	Attends private screening of Charlie Chaplin's *The Pilgrim*.
	Afternoon sailing party.
	Reception at the Asociación Politécnica.
	Banquet hosted by the University of the Republic at the Hotel del Prado. Awarded honorary professorship of the university.
May 1	Leaves Montevideo on board the French ship *Valdivia*.
May 4	Arrival in Rio de Janeiro. Lodges at the Hotel Glória.
May 5	Strolls with Isidoro Kohn, president of the Jewish community.

	Afternoon meeting with local German community, hosted by German merchants.
	Receives honorary doctorate from University of Rio de Janeiro.
	Outing to Sugarloaf Mountain with group of professors.
	Reception hosted by Jewish associations.
May 6	Stroll with Professor of Medicine Antônio da Silva Mello in the Santa Teresa neighborhood.
	Audiences with the president of the Republic Artur Bernardes, ministers, and Mayor Alaor Prata Soares.
	First public lecture on relativity at the Engineering Club.
May 7	Visit at National Museum of Brazil.
	Luncheon hosted by head of Faculty of Medicine, Aloísio de Castro.
	Session in Einstein's honor at the Brazilian Academy of Sciences. Lecture on the current state of the theory of light. Confirmed as corresponding member.
	Recording of short address at Rádio Sociedade (Text 27 in Additional Texts section).
May 8	Visit at the Oswaldo Cruz Institute.
	Second public lecture on relativity in the afternoon at the Polytechnic School.

	Dinner hosted by German ambassador Hubert Knipping at the Club "Germania."
May 9	Visit to the National Observatory.
	Lunch at residence of Antônio da Silva Mello.
	Visit with physiologists Álvaro and Miguel Ozório de Almeida.
	Dinner with the Kohns.
	Reception by Jewish community at the Automobile Club of Brazil.
May 10	Tour of neighborhoods and environs of Rio de Janeiro. Funicular to Corcovado.
	Evening receptions at Zionist Center and Scholem Aleichem Library.
May 11	Tour of the National Hospital for the Insane.
	Visits Brazilian ministers.
	Visits Brazilian Press Association. Screening of film on General Cândido Rondon.
	Dinner hosted by German ambassador at the Hotel Glória.
May 12	Departs Rio de Janeiro for Europe on board the S.S. *Cap Norte*.
May 27	Docks at Bilbao.
May 31	Disembarks in Hamburg.
June 1	Returns to Berlin.

Abbreviations

Descriptive Codes

AD	Autograph Document
AKS	Autograph Postcard Signed
ALS	Autograph Letter Signed
ALSX	Autograph Letter Signed Xerox
ALX	Autograph Letter Xerox
Tgm	Telegram
TLS	Typed Letter Signed
TTrD	Typed Transcript Document

Location Symbols

AEA	Albert Einstein Archives, Hebrew University of Jerusalem
CaPsCA	Institute Archives, California Institute of Technology, Pasadena, California
GyBPAAA	Politisches Archiv des Auswärtigen Amtes, Berlin
IsJNLI	National Library of Israel, Jerusalem
NoONPPC	Nobel Peace Prize Committee, Oslo

Notes

Preface

1. See Einstein to Hans Albert Einstein, 17 October 1918 [*CPAE 1998*, Vol. 8, Doc. 634].
2. See Einstein to Pauline Einstein, 8 October 1918 [*CPAE 1998*, Vol. 8, Doc. 631].
3. See Einstein to Paul and Maja Winteler-Einstein, and Pauline Einstein, 23 September 1918 [*CPAE 1998*, Vol. 8, Doc. 621].
4. See Einstein to Mileva Einstein-Marić, ca. 9 November 1918 [*CPAE 1998*, Vol. 8, Doc. 647].

Historical Introduction

1. For general overviews of the trip to South America, see *Grundmann 2004*, pp. 256–264, *Eisinger 2011*, pp. 73–94, and *Calaprice et al. 2015*, pp. 116–118.
2. See "South American Travel Diary Argentina, Uruguay, Brazil, 5 March–11 May 1925 [*CPAE 2015*, Vol. 14, Doc. 455, pp. 688–708].
3. See *Tolmasquim 2003*, pp. 177–198.
4. See *Moraes 2019*, Appendix I, pp. 204–208.
5. On Einstein's trip to the United States in the spring of 1921, see *CPAE 2009*, Vol. 12, Introduction, pp. xxviii–xxxviii.
6. See *Einstein 2018* for the published version of the travel diary to the Far East, Palestine, and Spain; "Amerika-Reise 1930," 30 November 1930–15 June 1931 [AEA, 29 134]; "Travel diary for USA," 3 December 1931–4 February 1932 [AEA, 29 136]; "Reise nach Pasadena XII 1932," 10 December 1932–18 December 1932 [AEA, 29 138]; and "Travel Diary for Pasadena," 28 January 1933–16 February 1933 [AEA, 29 143].
7. See Text 18 in the Additional Texts section of this volume.
8. See *Sayen 1985*, p. 72.
9. See *Calaprice et al. 2015*, p. 10.
10. See Einstein to Hans Albert and Eduard Einstein, 17 December 1922 in *Einstein 2018*, p. 252.
11. For discussions on the various factors, see *Ortiz 1995*, pp. 68–69 and 81–84, *Glick 1999*, pp. 102–106, *Ortiz and Otero 2001*, pp. 1–3, *Tolmasquim 2003*, pp. 56–63, *Grundmann 2004*, pp. 256–258, *Asúa and Hurtado de Mendoza 2006*, pp. 98–104, *Tolmasquim 2012*, pp. 121–123, and *Gangui and Ortiz 2014*, pp. 167–175.
12. See *Ortiz 1995*, pp. 67–68.
13. See ibid, p. 77.

14. See ibid, pp. 68 and 78, and Eduardo L. Ortiz. "The emergence of theoretical physics in Argentina, Mathematics, mathematical physics and theoretical physics 1900–1950." Accessed 4 June 2021. *Proceedings of Science* (Héctor Rubinstein Memorial Symposium, 2010) 030. https://pos.sissa.it/109/030/pdf, p. 9.
15. See *Gangui and Ortiz 2014*, p. 168.
16. These included the prominent engineer Enrique Butty, the German-educated physicists José Collo and Teófilo Isnardi, and the astronomer Félix Aguilar (see *Asúa and Hurtado de Mendoza 2006*, p. 100).
17. See *Paty 1999*, p. 349.
18. See *Tolmasquim 2012*, p. 121.
19. See *Moreira 1995*, p. 178. Interestingly, both Argentina and Brazil were involved in attempts to verify general relativity. The National Observatory of Córdoba participated in the unsuccessful expeditions to observe the solar eclipses in 1912 and 1914 and the National Observatory of Rio de Janeiro played a crucial role in the successful eclipse expedition to Sobral in the Brazilian state of Ceará in 1919 (see Richard A. Campos. "Still Shrouded in Mystery: The Photon in 1925." Accessed 3 June 2021. https://www.yumpu.com/en/document/read/5228000/physics-0401044-pdf-arxiv, p. 9).
20. See *Tolmasquim 2003*, pp. 111–112.
21. See *Moreira 1995*, pp. 177 and 181.
22. See *Tolmasquim and Moreira 2002*, p. 232.
23. See *Tolmasquim 2003*, pp. 118.
24. See Campos. "Still Shrouded in Mystery: The Photon in 1925," p. 9.
25. See *CPAE 2009*, Vol. 12, Calendar, entry for 3 August 1921; Erich Wende to Einstein, 11 October 1921 [*CPAE 2009*, Vol. 12, Doc. 264]; and Einstein to Erich Wende, 13 October 1921 [*CPAE 2009*, Vol. 12, Doc. 267].
26. See *Ortiz 1995*, p. 81.
27. See ibid, pp. 82 and 84; *Grundmann 2004*, pp. 256–257; and Report of the German Embassy in Buenos Aires to the German Foreign Ministry, Berlin, 26 September 1922 (GyBPAAA, R 64677).
28. See *Ortiz 1995*, p. 82.
29. See Joaquin J. Stutzin to Einstein, 30 October 1923 [*CPAE 2015*, Vol. 14, Abs. 198], and *Mundo Israelita*, 7 March 1925.
30. See Text 1 in the Additional Texts section of this volume.
31. See *Ortiz 1995*, p. 92.
32. See ibid, p. 94.
33. See ibid, pp. 82–84.
34. See Text 2 in the Additional Texts section of this volume.
35. See *Ortiz 1995*, p. 69.
36. See *Asúa and Hurtado de Mendoza 2006*, p. 102.
37. See *Ortiz 1995*, pp. 118–119.
38. See *Tolmasquim 2003*, pp. 56 and 58.
39. See *Raffalovich 1952*, pp. 300–301, and *Glick 1999*, p. 103.

40. See *Glick 1999*, p. 103.
41. See *Tolmasquim and Moreira 2002*, p. 233.
42. See Text 6 in the Additional Texts section of this volume.
43. See *Raffalovich 1952*, pp. 300–301, and *Glick 1999*, p. 104.
44. See *Grundmann 2004*, pp. 180–182.
45. See *Renn 2013*, p. 2577.
46. See Einstein to Fritz Haber, 6 October 1920 [*CPAE 2006*, Vol. 10, Doc. 162].
47. See *Gaviola 1952*, p. 238, and *Asúa and Hurtado de Mendoza 2006*, p. 103.
48. See Text 3 in the Additional Texts section of this volume.
49. See Text 4 in the Additional Texts section of this volume.
50. See Einstein to Hans Albert Einstein, 27 October 1924 [*CPAE 2015*, Vol. 14, Doc. 348].
51. See *Tolmasquim 2012*, p. 123.
52. See *Asúa and Hurtado de Mendoza 2006*, p. 93.
53. See *Einstein 2018*, p. 65.
54. See Einstein to Svante Arrhenius, 10 January 1923, in *Einstein 2018*, p. 258.
55. See Text 5 in the Additional Texts section of this volume.
56. See Text 7 in the Additional Texts section of this volume.
57. See Einstein to Robert A. Millikan, 3 November 1924 [*CPAE 2015*, Vol. 14, Doc. 306, p. 552], and *Tolmasquim 2003*, p. 59.
58. See *Tolmasquim 2003*, p. 63.
59. See ibid, p. 61.
60. See *Einstein 2018*, p. 72.
61. See *Tolmasquim 2003*, p. 63.
62. See ibid, p. 55.
63. See Einstein to Betty Neumann, 5 June 1924 [*CPAE 2015*, Vol. 4, Doc. 262, p. 401].
64. See *Fölsing 1997*, p. 548.
65. See Einstein to Flora Neumann-Mühsam, 11 March 1925 [*CPAE 2015*, Vol. 14, Doc. 459, p. 713].
66. On the history of this period, see *Romero 2013*, pp. 27–55, and *Hedges 2015*, pp. 24–45.
67. On the history of this period, see *Bethell 1986*, Vol. 5, pp. 453–474, and *Andrews 2010*, pp. 2–5.
68. On the history of the First Republic, see *MacLachlan 2003*, pp. 39–90, and *Fausto and Fausto 2014*, pp. 144–186.
69. See *Skidmore 1995*, p. 95.
70. See *Santos and Hallewell 2002*, p. 79.
71. See *Bethell 2010*, p. 461.
72. See *Fausto and Fausto 2014*, p. 184.
73. See text of diary, this volume, note 98.
74. See *Siebenmann 1988*, pp. 67 and 71, *Siebenmann 1992a*, p. 16, *Siebenmann 1992b*, p. 194, and *Minguet 1992*, pp. 107–108.
75. See *Minguet 1992*, p. 108.
76. See *Glick 1999*, p. 108.

77. See *Siebenmann 1988*, p. 66–70, *König 1992*, p. 227, and *Minguet 1992*, p. 115.
78. See *Siebenmann 1992b*, p. 199.
79. See *König 1992*, p. 210, and *Minguet 1992*, p. 115.
80. See *König 1992*, p. 223.
81. See *Zapata 1979*, p. 51.
82. See ibid, pp. 52–53, and *Siebenmann 1992b*, p. 193.
83. See *Siebenmann 1992b*, p. 193.
84. See ibid, p. 198.
85. See *Newton 1977*, p. 93.
86. See ibid, pp. 93–97 and 108.
87. See Julius Koch to Jakob, Fanny, and Pauline Koch, 26 June 1889 [AEA, 143 392].
88. See *Zapata 1979*, p. 52.
89. See *CPAE 1993*, Vol. 5, p. 640, and *CPAE 2015*, Vol. 14, p. 742.
90. See *CPAE 2009*, Vol. 12, p. 310, and the section "Einstein's Acceptance of the Invitations" in this introduction.
91. See text of diary, this volume, entry for 6 March 1925.
92. See ibid, entry for 19 March 1925. The phrase "member of the idle rich class" appears in the appendix, "The Revolutionist's Handbook," in *Shaw 1919*.
93. See text of diary, this volume, entry for 24 March 1925.
94. See ibid.
95. See Text 12 in the Additional Texts section of this volume.
96. See ibid.
97. See text of diary, this volume, entry for 24 March 1925.
98. See ibid, entry for 26 March 1925.
99. See Text 12 in the Additional Texts section of this volume.
100. See text of diary, this volume, entry for 24 March 1925.
101. See ibid, entry for 27 March 1925.
102. See ibid, entry for 28 March 1925.
103. See Text 14 in the Additional Texts section of this volume.
104. See ibid.
105. See Text 18 in the Additional Texts section of this volume.
106. See Text 14 in the Additional Texts section of this volume.
107. See text of diary, this volume, entry for 14 April 1925.
108. See ibid.
109. See Text 18 in the Additional Texts section of this volume.
110. See text of diary, this volume, entry for 21 April 1925.
111. See ibid, entry for 6 April 1925.
112. See ibid, entry for 7 April 1925.
113. See ibid, entry for 14 April 1925.
114. See ibid, entry for 1 April 1925.
115. See ibid, entry for 12 April 1925
116. See ibid, entry for 14 April 1925.

117. See ibid, entry for 13 April 1925.
118. See ibid, entry for 2 April 1925.
119. See ibid, entry for 19 March 1925.
120. See Text 18 in the Additional Texts section of this volume.
121. See Text 31 in the Additional Texts section of this volume.
122. See Text 34 in the Additional Texts section of this volume.
123. See Einstein to Michele Besso, before 26 July 1920 [*CPAE 2006*, Vol. 10, Doc. 85, p. 346].
124. See Einstein to Max Planck, 6 July 1922 [*CPAE 2012*, Vol. 13, Doc. 266, p. 392], and *Einstein 2018*, p. 18.
125. See Einstein to Heinrich Zangger, 16 February 1917 [*CPAE 2006*, Vol. 8, Doc. 299a, in Vol. 10, p. 72], and Einstein to Heinrich Zangger, 18 June 1922 [*CPAE 2012*, Vol. 13, Doc. 241, p. 370].
126. See *Einstein 2018*, p. 40.
127. See Text 7 in the Additional Texts section of this volume.
128. See text of diary, this volume, entry for 14 April 1925.
129. See ibid, entry for 24 April 1925.
130. See ibid, entry for 26 April 1925.
131. See ibid, entries for 26 and 30 April 1925.
132. See Text 22 in the Additional Texts section of this volume.
133. See text of diary, this volume, entry for 30 April 1925.
134. See ibid, entry for 24 April 1925.
135. See *Frank 1947*, p. 100.
136. See Einstein to Pauline Einstein, 16 June 1919 [*CPAE 2004*, Vol. 9, Doc. 61, p. 91], and *Einstein 1917*, 1920 appendix [*CPAE 1996*, Vol. 6, Doc. 42, p. 512].
137. See text of the diary, this volume, entry for 22 March 1925.
138. See Text 12 in the Additional Texts section of this volume.
139. See *Revista Fon-Fon* 13 (28 March 1925): 50.
140. See text of the diary, this volume, entry for 4 May 1925.
141. See ibid, entry for 5 May 1925.
142. See Text 34 in the Additional Texts section of this volume.
143. See text of the diary, this volume, entry for 4 May 1925.
144. See ibid, entry for 6 May 1925.
145. See ibid, entry for 7 May 1925.
146. See ibid, entry for 5 May 1925.
147. See ibid.
148. See text of the diary, this volume, entries for 6 and 9 May 1925.
149. See ibid, entry for 7 May 1925.
150. See ibid, entry for 11 May 1925.
151. See ibid, entry for 6 May 1925.
152. See ibid, entry for 9 May 1925.
153. See *Illustração Brasileira* 58 (June 1925): 50.
154. See text of the diary, this volume, entry for 7 May 1925.

155. See Text 25 in the Additional Texts section of this volume.
156. See Text 29 in the Additional Texts section of this volume.
157. In his Far East travel diary, Einstein included the following comment on the Sinhalese inhabitants: "For all their fineness, they give the impression that the climate prevents them from thinking backward or forward by more than a quarter of an hour" (see *Einstein 2018*, p. 105).
158. See *Einstein 2018*, pp. 53–59.
159. See ibid, p. 42.
160. See *Liebman 1973*, p. 9.
161. See *Mirelman 1990*, p. 168.
162. See *Rosenswaike 1960*, pp. 202–203.
163. See *Liebman 1973*, p. 8.
164. See *Horowitz 1962*, p. 208, *Avni 1972*, p. 266, and *Liebman 1973*, p. 12.
165. See *Levine 1987*, p. 73.
166. See *Schenkolewski-Kroll 2017*, p. 80.
167. See text of the diary, this volume, entry for 31 March 1925.
168. See *Rosenkranz 2011*, p. 255.
169. See text of the diary, this volume, entry for 26 March 1925.
170. See ibid, entry for 6 April 1925.
171. See ibid, entry for 16 April 1925.
172. See ibid, entry for 21 April 1925.
173. See Text 20 in the Additional Texts section of this volume.
174. See text of the diary, this volume, entry for 24 March 1925.
175. See ibid, entry for 21 April 1925.
176. See Texts 13, 15, and 19 in the Additional Texts section of this volume.
177. See Text 20 in the Additional Texts section of this volume.
178. See *Sapolinsky 1963*, pp. 74–75, 81–82, and 86.
179. See *El Día*, 24 April 1925.
180. See text of diary, this volume, entry for 24 April 1925.
181. See ibid, note 77, and entry for 28 April 1925.
182. See *Elkin 2014*, p. 259.
183. See *Lesser 1989*, pp. 4–6 and 10–11.
184. See ibid, p. 14.
185. See ibid, p. 118.
186. See *Gherman 2017*, p. 197.
187. See text of diary, this volume, entry for 9 May 1925, and note 101.
188. See ibid, entry for 10 May 1925, and note 102.
189. See Text 32 in the Additional Texts section of this volume.
190. See Text 34 in the Additional Texts section of this volume.
191. See *Tolmasquim 2012*, p. 125.
192. See *Bernecker and Fischer 1992*, p. 198.
193. See *Newton 1977*, pp. 103 and 105.

194. See *Werner 1996*, p. 185.
195. See *Newton 1977*, p. 124.
196. See ibid, p. 112 and 146.
197. See *Tolmasquim 2012*, p. 126.
198. See ibid.
199. See Text 18 in the Additonal Texts section of this volume.
200. See text of diary, this volume, entry for 17 April 1925.
201. See *Die Deutsche La Plata Zeitung*, 18 April 1925.
202. See German Embassy, Buenos Aires, to German Foreign Ministry, Berlin, 30 April 1925 (GyBPAAA, R 64678).
203. See "Ausstellung zu 160 Jahre diplomatische Beziehungen zu Uruguay." 15 July 2016. Accessed 3 May 2021. https://www.deutschland.de/en/node/3639.
204. See Walther L. Bernecker. "Siedlungskolonien und Elitenwanderung. Deutsche in Lateinamerika: das 19. Jahrhundert." Accessed 3 May 2021. https://www.matices-magazin.de/archiv/15-deutsche-in-lateinamerika/deutsche-in-lateinamerika/, and Torsten Eßer. "Deutsche in Lateinamerika." Accessed 3 May 2021. http://www.torstenesser.de/download-text/Deutsche%20in%20Lateinamerika.pdf.
205. See *Beretta Curi 2018*, pp. 81–82.
206. See "Deutsch-uruguayische Beziehungen." Accessed 3 May 2021. https://www.pangloss.de/cms/index.php?page=uruguay, and *Schonebaum 1998*, p. 237.
207. See text of diary, this volume, entry for 28 April 1925.
208. See German Embassy, Montevideo, to the German Foreign Ministry, Berlin, 4 June 1925 (GyBPAAA, R 64678).
209. See Text 22 in the Additional Texts section of this volume.
210. See text of diary, this volume, entry for 29 April 1925.
211. See ibid, entry for 3 May 1925.
212. See *Oberacker 1979*, pp. 222, 224, and 238.
213. See text of diary, this volume, entry for 5 May 1925.
214. See ibid, entry for 8 May 1925.
215. See Report of the German Embassy, Rio de Janeiro, 20 May 1925 (GyBPAAA, R 64678).
216. See Carl Gneist, German Embassy in Buenos Aires, to German Foreign Ministry, Berlin, 30 April 1925 (GyBPAAA, R 64678).
217. See German Embassy, Montevideo, to the German Foreign Ministry, Berlin, 4 June 1925 (GyBPAAA, R64678).
218. See Germany Embassy, Rio de Janeiro, to the German Foreign Ministry, Berlin, 20 May 1925 (GyBPAAA, R 64678).
219. See Einstein to Paul Ehrenfest, 22 March 1919 [*CPAE 2004*, Vol. 9, Doc. 10].
220. See *Einstein 2018*, p. 41.
221. See Text 14 in the Additional Texts section of this volume.
222. See text of diary, this volume, entry for 24 April 1925.
223. See ibid, entry for 4 May 1925.

224. See Text 31 in the Additional Texts section of this volume.
225. See Text 32 in the Additional Texts section of this volume.
226. See Text 23 in the Additional Texts section of this volume.
227. See "Een Interview met Prof. Albert Einstein," *Nieuwe Rotterdamsche Courant*, 4 July 1921, and "Einsteins amerikanische Eindrücke. Was er wirklich sah," *Vossische Zeitung*, 10 July 1921.
228. See *Einstein 2018*, pp. 48–50 and *Kaplan 1997*, p. 22.
229. See text of the diary, this volume, entry for 6 May 1925.
230. See *Kaplan 1997*, p. 6.
231. See ibid, p. 22.
232. See text of diary, this volume, entry for 8 March 1925.
233. See ibid, entry for 11 March 1925.
234. See ibid, entry for 8 March 1925.
235. See ibid, entry for 9 March 1925.
236. See ibid, entry for 14 March 1925.
237. See ibid, entry for 19 March 1925.
238. See ibid, entry for 24 March 1925.
239. See ibid.
240. See ibid.
241. See text of diary, this volume, entry for 1 April 1925.
242. See ibid, entry for 18 April 1925.
243. See ibid, entry for 21 April 1925.
244. See Text 11 in the Additional Texts section of this volume.
245. See Text 20 in the Additional Texts section of this volume.
246. See Text 18 in the Additional Texts section of this volume.
247. See *Mudimbe-Boyi 1992*, p. 28.
248. See *Seth and Knox 2006*, pp. 4–5, 214.
249. See ibid, p. 6.
250. See *Mudimbe-Boyi 1992*, p. 27.
251. See Text 5 in the Additional Texts section of this volume.
252. See Text 9 in the Additional Texts section of this volume.
253. See text of diary, this volume, entry for 5 March 1925.
254. See ibid, entry for 17 March 1925.
255. See Text 12 in the Additional Texts section of this volume.
256. See text of diary, this volume, entry for 13 April 1925.
257. See ibid, entry for 14 April 1925.
258. See ibid, entry for 6 May 1925.
259. See ibid, entry for 18 April 1925.
260. See ibid, entry for 26 April 1925.
261. See Text 20 in the Additional Texts section of this volume.
262. See Text 12 in the Additional Texts section of this volume.
263. See text of diary, this volume, entry for 3 May 1925.

264. See ibid, entries for 1 and 3 May 1925.
265. See ibid, entry for 17 April 1925.
266. See *Leerssen 2000*, pp. 280–284.
267. See *Einstein 2018*, p. 58. Einstein's attitudes on race did not evolve in a vacuum. For a discussion of the broader historical contexts in which the views of European, German, and Jewish intellectuals on race evolved, see *Einstein 2018*, pp. 53–55.
268. See *Miles and Brown 2003*, p. 104.
269. See ibid, p. 103.
270. See text of the diary, this volume, entry for 7 May 1925.
271. See Text 14 in the Additional Texts section of this volume.
272. See Text 18 in the Additional Texts section of this volume.
273. See text of the diary, this volume, entry for 22 March 1925.
274. See ibid, entry for 11 May 1925.
275. See *Skidmore 1995*, p. 92.
276. See ibid, p. 95.
277. See *Einstein 2018*, pp. 60 and 65.
278. See *Walton 2009*, p. 117.
279. See *Einstein 2018*, pp. 60–62.
280. See *CPAE 2015*, Vol. 14, Introduction, p. lxiv.
281. See *Grundmann 2004*, p. 258.
282. See text of diary, this volume, entry for 17 March 1925.
283. See Text 11 in the Additional Texts section of this volume.
284. See ibid.
285. See text of diary, this volume, entry for 14 April 1925.
286. See ibid, entry for 9 May 1925.
287. See ibid, entry for 26 March 1925.
288. See ibid, entry for 11 May 1925.
289. See ibid, entry for 7 March 1925.
290. See Text 9 in the Additional Texts section of this volume.
291. See text of diary, this volume, entries for 12, 17, and 19 March 1925.
292. See Text 11 in the Additional Texts section of this volume.
293. See Text 18 in the Additional Texts section of this volume.
294. See *Einstein 2018*, p. 329, note 34.
295. See text of diary, this volume, entry for 2 May 1925.
296. See *Einstein 1923a*.
297. See *Einstein 1925*.
298. See *Ortiz 1995*, p. 122.
299. See *Gangui and Ortiz 2014*, p. 185.
300. See *Asúa and Hurtado de Mendoza 2006*, pp. 119 and 121.
301. See *Gangui and Ortiz 2014*, p. 185.
302. See *Asúa and Hurtado de Mendoza 2006*, pp. 108 and 121.
303. See *Ortiz 1995*, p. 105.

304. See Eduardo L. Ortiz. "The emergence of theoretical physics in Argentina, Mathematics, mathematical physics and theoretical physics 1900–1950." Accessed 4 June 2021. *Proceedings of Science* (Héctor Rubinstein Memorial Symposium, 2010) 030. https://pos.sissa.it/109/030/pdf, p. 10.
305. See *Ortiz and Otero 2001*, p. 34.
306. See Ortiz. "The emergence of theoretical physics in Argentina, Mathematics, mathematical physics and theoretical physics 1900–1950," p. 10.
307. See *Gangui and Ortiz 2014*, p. 185.
308. See *Ortiz 1995*, p. 122.
309. See *Gangui and Ortiz 2014*, p. 185.
310. See ibid.
311. See *Ortiz 1995*, pp. 116–117 and *Gangui and Ortiz 2014*, p. 186.
312. See *Ortiz and Otero 2001*, p. 10.
313. See ibid, pp. 11 and 22.
314. See ibid, p. 21.
315. See ibid, p. 25.
316. See ibid, p. 9.
317. See *Ortiz 1995*, pp. 124–125.
318. See *Ortiz and Otero 2001*, p. 22.
319. See *Tolmasquim 2003*, pp. 163–164.
320. See *Moreira 1995*, p. 195, and *Glick 1999*, p. 114.
321. See *Glick 1999*, p. 106, and *Tolmasquim 2003*, p. 163.
322. See *Tolmasquim and Moreira 2002*, p. 236.
323. See *Glick 1999*, p. 113.
324. See *Moreira 1995*, p. 186, and *Silva da Silva 2005*, pp. 294–296.
325. See *Moreira 1995*, pp. 188–189.
326. See *Moreira 1995*, p. 190, and *Tolmasquim and Moreira 2002*, p. 236.
327. See *Moreira 1995*, p. 196, and *Glick 1999*, p. 113.
328. Einstein would frequently stay at a bachelor's apartment that had been refurbished for him at the home and factory of his friend, the industrialist Hermann Anschütz-Kaempfe, in Kiel and at the home of his close friend Paul Ehrenfest in Leyden. He repeatedly referred to life in the German capital as "nerve-wracking" (see, e.g., Einstein to Elsa Einstein, 14 September 1920 [*CPAE 2006*, Vol. 10, Doc. 149]).
329. See Text 14 in the Additional Texts section of this volume.
330. See Texts 18 and 22 in the Additional Texts section of this volume.
331. See text of diary, this volume, entry for 19 March 1925.
332. See, e.g., Einstein to Hans Albert Einstein, 18 June 1921 [*CPAE 2009*, Vol. 12, Doc. 153] in which he planned to purchase third-class railway tickets.
333. See *Einstein 2018*, p. 288, note 308.
334. On Einstein's illness of 1928, see *CPAE 2021*, Vol. 16, Introduction, pp. lv-lx.
335. See *Einstein 2018*, p. 189.
336. See text of the diary, this volume, entry for 11 May 1925.

Travel Diary: Argentina, Uruguay, Brazil, 5 March–11 May 1925

1. AD. [AEA, 29 132]. Published as "South American Travel Diary Argentina, Uruguay, Brazil" in *CPAE 2015*, Vol. 14, Doc. 455, pp. 688–708. The document presented here is preserved in a notebook that measures 16.3 x 10.2 cm and consists of 72 lined pages. The notebook includes 43 lined pages of travel diary entries, followed by 29 blank lined pages. The pages of the notebook have been numbered by the Albert Einstein Archives. On the notebook cover, Einstein wrote the title "TAGEBUCH" (Diary) and it seems that a sketched drawing beneath the title was added by hand. Einstein's long-time secretary Helen Dukas has noted "Südamerika 1925" ("South America 1925") above the title. A sticker below the drawing reads "Diary trip South-America, 1925" (in type) and "item 2—cardfile & typescr." in Helen Dukas's hand. The diary entries appear on pp. 2 to 43 and are written in ink. The notebook pages, which may have originally been stapled together, are sewn together with white yarn. A TTrD with Helen Dukas's explanatory notes is also available [AEA, 5 255]. These notes have not been included.
2. Elsa Einstein (1876–1936), his second wife. Moritz Katzenstein (1872–1932) was director of the Second Surgical Department at the Municipal Hospital in Friedrichshain, Berlin, and one of Einstein's close friends. He had five sisters: Henriette, Rosa, Sara, Emma, and Paula. Alexander Bärwald (1877–1930) was a German–Jewish architect who designed the Technion building in Haifa. His wife was Charlotte Bärwald-Eisenberg (1883–1937).
3. In Hamburg.
4. Probably Emily Robinow-Kukla (1883–1967) who was the wife of Hamburg Jewish merchant Paul Robinow. Her son-in-law was Wilhelm Roloff (1899–1949), a German physician at the Charité Hospital in Berlin.
5. Carl Melchior (1871–1933), who was a Hamburg Jewish banker, a partner at M. M. Warburg & Co., and chairman of the Supervisory Board of Beiersdorf AG, the skin-care company that invented the Nivea Creme.
6. The S.S. *Cap Polonio* was owned by the Hamburg South American Line.
7. The Robinows' son-in-law was Wilhelm Roloff.
8. 23,000 tons. Einstein's younger step-daughter Margot Einstein (1899–1986) had apparently been planning to accompany him on his trip, but had presumably fallen ill shortly before the planned departure (see Text 9 in the Additional Texts section of this volume).
9. Carl Jesinghaus (1886–1948) was Professor of Philosophy and Psychology at the University of Buenos Aires.
10. Ideas about the foundations of Riemannian geometry, the mathematical framework for the general theory of relativity, had previously been the starting point for approaches to establish a unified field theory of the electromagnetic and gravitational fields. One such approach was following up upon the work of Hermann Weyl and

Arthur S. Eddington, and based Riemannian geometry on the concept of the affine connection. On his return trip from Japan in early 1923, such an idea had led to Einstein's exploration of this approach, resulting in the publication of *Einstein 1923a* (*CPAE 2012*, Vol. 13, Doc. 425), *1923b*, and *1923c* (*CPAE 2015*, Vol. 14, Docs. 13 and 52). By mid-1923, Einstein had abandoned this approach (see *CPAE 2015*, Vol. 14, Introduction, pp. xxxvii–xl). Later in 1925, however, he would publish on a new approach to a unification of the fields based on the assumption of an asymmetric metric; see *Einstein 1925*, presented to the Prussian Academy on 9 July 1925.

11. *Chaucer 1924*.
12. Else Jerusalem-Kotányi (1876–1943) was an Austrian-born Jewish writer who had been living in Buenos Aires since 1911. Her most well-known works challenged bourgeois notions of morality and sexuality (see *Spreitzer 2016*).
13. A Coruña and Vigo are cities in Galicia in northwest Spain.
14. Else Jerusalem.
15. Carl Jesinghaus. Possibly Johannes Sievers (1880–1969), art historian and legation councilor at the German Foreign Ministry.
16. Probably the Castelo de São Jorge.
17. The Mosteiro dos Jerónimos.
18. *Meyerson 1925*.
19. For an analysis of this equation, see "South American Travel Diary Argentina, Uruguay, Brazil" in *CPAE 2015*, Vol. 14, Doc. 455, note 15.
20. *Koigen 1925*, an autobiographical work in which the Ukrainian–Jewish author described the years he spent in Kyiv, his family's escape to Germany, and the traumas he suffered due to pogroms and violence in the Russian Civil War.
21. Probably *Jerusalem 1928*, in which a young rabbi named David is doomed, not only by the clash between tradition and modernity, but also by the wrathful actions of the God of the Old Testament (see *Stürzer 1993*, p. 378).
22. Einstein turned 46 on 14 March. He thanks Elsa for her birthday letter in Text 10 in the Additional Texts section of this volume.
23. The equivalent of 28°C and 84°F.
24. For a historical discussion of these field equations in the context of Einstein's attempts at a field-theoretic description of matter, see *Sauer 2012*. For earlier doubts about the viability of field theory, see *Einstein 1924*, §1.
25. Wolfgang Amadeus Mozart, "Eine kleine Nachtmusik," KV 525, and Ludwig van Beethoven, "Romanze" No. 2, F major, op. 50.
26. The phrase "member of the idle rich class" appears in the appendix, "The Revolutionist's Handbook," to *Shaw 1919*. In the original of the diary, Einstein uses the abbreviation "M. d. R. F." for "Mitglied der reichen Faulenzerklasse," the German translation of MIRC.
27. This pertains to Niels Bohr's theory of light emission that led to the problem of how to explain that the emitted light quantum appears to travel as a directed particle,

whereas the emitted light could also show interference phenomena. For more details, see "South American Travel Diary Argentina, Uruguay, Brazil" in *CPAE 2015*, Vol. 14, Doc. 455, note 23.

28. The large delegation that greeted Einstein at the port consisted of prominent members of the local Jewish community, including Rabbi Isaiah Raffalovich (1870–1956), spiritual leader of the Jewish community of Rio de Janeiro and director of the Jewish Colonization Association in Brazil, and Isidoro Kohn (1877–1965), an Austrian-born businessman and president of the Jewish community. The scientific welcoming committee was headed by the astronomer Henrique Morize and included the mathematicians Ignácio do Amaral and Roberto Marinho de Azevedo; Paulo de Frontin, director of the Polytechnic School; Aloísio de Castro (1881–1959), chairman of the Faculty of Medicine at the University of Rio de Janeiro and member of the International Committee on Intellectual Cooperation; and the German-born chemist Daniel Henninger. Numerous journalists were also in attendance.

    The welcoming committee invited Einstein to tour the city by car. In the convoy of seven cars, Einstein was accompanied by Rabbi Raffalovich (see *Tolmasquim 2003*, pp. 70 and 232, and *Glick 1999*, pp. 104–105).
29. Einstein was greeted by the director of the Botanical Garden of Rio de Janeiro, Antonio Pacheco Leão, and entered a note in the visitors' log. Following the tour of the garden, Einstein lunched at the recently constructed Copacabana Palace Hotel, hosted by Assis Chateaubriand, owner of the newspaper *O Jornal*. Other guests included Eduardo Horovitz, secretary of the Zionist Federation; Leon Schwartz, president of the Hebrew College; and Emanuel Galano, president of the Sephardi synagogue Bnei Herzl. Following lunch, Einstein went for a walk (see *O Jornal*, *O Paiz* and *A Noite*, 21 and 22 March 1925 and *Glick 1999*, p. 104).
30. Mauricio Nirenstein (1877–1935) was Professor of Spanish Literature and secretary-general of the National University of Buenos Aires. A small group of Argentinian academics headed by Nirenstein came to greet Einstein upon his arrival in Montevideo and travel with him across the River Plate to Buenos Aires. The other prominent member of the group was the civil engineer Enrique Butty (*Ortiz 1995*, p. 99).
31. For a comprehensive list of the Jewish organizations represented at the harbor to greet Einstein, see *Ortiz 1995*, p. 107, note 125. A joint Jewish reception committee was established whose members were Sansón Raskowsky, David Groisman, León Horischnik, L. Minuchin, and Luis Sverdlick (see *Ortiz 1995*, p. 108).
32. Bruno Wassermann (1874–1940) was a German-Jewish paper merchant. The Wassermanns' house was located in the elegant neighborhood of Belgrano.
33. The German ambassador was Carl Gneist (1868–1939). Leopoldo Lugones visited Einstein on his first day in Buenos Aires (see *Ortiz 1995*, p. 100).
34. The rector was José Arce (1881–1968), Professor of Anatomy and Surgery and a politician. The dean of the Faculty of Exact, Physical, and Natural Sciences was Eduardo Huergo (1873–1929), Professor of Civil Engineering.

35. Einstein gave extensive interviews to reporters of *La Nación* and *Crítica*. During his tour of the city, he visited the parks of the Palermo neighborhood and the fruit and vegetable market at the Abasto (*Ortiz 1995*, p. 100).
36. Julio R. Castiñeiras (1885–1944) and his wife, Delia O. Miguel de Castiñeiras.
37. In the original, Einstein uses a Berlin dialect phrase for "I've had enough."
38. During his tour of the United States, Einstein had participated in numerous events hosted by Jewish organizations and had subsequently expressed some disenchantment with his fellow Jews (see *CPAE 2009*, Vol. 12, Introduction, pp. xxxii–xxxiv, and Einstein to Michele Besso, before 30 May 1921 [*CPAE 2009*, Vol. 12, Doc. 141]).
39. The reception and introductory public lecture were held on 27 March at the main auditorium of the Colegio Nacional de Buenos Aires, a prestigious high school affiliated with the university. The lecture lasted approximately half an hour and dealt with inertial reference systems, the physical experiments performed to verify the principle of relativity, and the problems raised by concepts of absolute space and ether. The lecture was delivered to a "wide and heterogeneous university audience" that included government ministers, foreign ambassadors, teachers, and students. At the podium were Minister of Justice and Education Antonio Sagarna (1874–1949), Foreign Minister Ángel Gallardo, and senior university officials (see *La Prensa*, 28 March 1925; *La Nación*, 28 March 1925; and *Gangui and Ortiz 2008*, p. 440).
40. Einstein's first scientific lecture took place on 28 March at the University of Buenos Aires' Faculty of Exact, Physical, and Natural Sciences, and dealt with the physical experiments performed to test the existence of the ether and the theories of Dutch physicist Hendrik A. Lorentz and Irish physicist George FitzGerald to overcome the difficulties encountered in those experiments, the work of American physicist Albert A. Michelson, and Einstein's own theory on the constancy of the speed of light. Originally, Einstein had planned to deliver twelve lectures at the University of Buenos Aires. However, their number was reduced to eight due to demand for lectures by other universities (see *La Nación*, 29 March 1925; *Ortiz 1995*, pp. 101 and 115; and *Gangui and Ortiz 2008*, p. 441). The minister of education was Antonio Sagarna.
41. Alfredo Hirsch (1872–1956) was one of the proprietors of the grain-exporting company Bunge y Born, a philanthropist and patron of the arts. His wife was Lisa Hirsch-Gottschalk. The reference to "Alice" pertains to Einstein's cousin Alice Steinhardt-Koch (1893–1975).
42. Einstein's cousin Robert Koch (1879–1952?). Einstein was 46 years old.
43. Einstein visited the editorial offices of *La Prensa* accompanied by Bruno Wassermann and Robert Koch. For press coverage of Einstein's visit, see *La Prensa*, 31 March 1925.
44. A "schul" is Yiddish for synagogue. Einstein visited an orphanage for girls, the Asilo Argentino de Huérfanas Israelitas (*Tolmasquim 2012*, p. 124).
45. Berta Wassermann-Bornberg (1878–1932) with whom Einstein was lodging in Buenos Aires. Einstein and Wassermann participated in the second flight of a Junkers hydroplane visiting the capital that performed maneuvers at a high altitude over the

city. The German pilot's name was Grundtker. The passengers of the first flight were German admiral Paul Behncke and his adjutant (see *La Prensa*, 2 April 1925).

46. The president of Argentina was Marcelo T. de Alvear (1868–1942). Leopoldo Lugones (1874–1938) was a prominent Argentinian writer, amateur scientist, and member of the International Committee on Intellectual Cooperation.
47. La Plata is located 58 kilometers southeast of Buenos Aires. Einstein was accompanied on his train trip to La Plata by the rector of the University of Buenos Aires, José Arce; the dean of the Faculty of Humanities and Education Sciences, Enrique Mouchet; and the physicist Ramón G. Loyarte. A luncheon was hosted by the president of the University of La Plata, Benito Nazar Anchorena, at the city's Jockey Club. The fall semester of the National University of La Plata was inaugurated by its president. A scientific session in Einstein's honor was presided over by Richard Gans, director of the university's Physics Institute. Einstein also met with members of a delegation of the Ateneo Juventud Israelita (see *La Prensa*, 3 April 1925, and *Ortiz 1995*, p. 101).
48. The luncheon, hosted by rector José Arce on behalf of the University of Buenos Aires, took place at the Elisabeth Hall of the Jockey Club and was attended by Minister Sagarna, Ambassador Gneist, and Carl Egger, the Swiss ambassador (see *La Prensa*, 4 April 1925).
49. Louis Louis-Dreyfus (1867–1940) was a French industrialist and employer of Robert Koch (see Text 14 in the Additional Texts section of this volume).

    Einstein delivered a brief lecture at the University of Buenos Aires' Faculty of Philosophy and Letters on the relation between geometry and relativity (see *La Nación*, 5 April 1925, and *Ortiz 1995*, p. 105).
50. The Wassermanns' country *estancia* was located in the town of Llavallol, 30 kilometers south of Buenos Aires.
51. The physiologist was Bernardo A. Houssay (1887–1971), a specialist in endocrinology and director of the Institute of Physiology at the University of Buenos Aires. Eugenio Pablo Fortin (1876–1947) was a French-born ophthalmologist and research scientist.

    The celebration of the official opening of the Hebrew University (which had taken place on 1 April on Mount Scopus in Jerusalem) was held at the city's largest theater, the Teatro Coliseo. The event was organized by the Zionist Federation of Argentina and presided over by its president, Isaac Nissensohn. The audience numbered some 4,000 people. Messages were read from the vice president of the Republic and the president of the Senate. For Einstein's speech, see Text 15 in the Additional Texts section of this volume. Ben-Zion Mossinson (1878–1942) was a member of the General Zionist Council. He was visiting Argentina as a representative of the Palestine Foundation Fund (see also *La Prensa*, 7 April 1925).
52. José Arce.
53. The three-day rest in Llavallol took place over the Easter break (see *Ortiz 1995*, p. 105). Einstein was interested in a theory that unified the gravitational and electromagnetic fields; see also note 10 above.

54. Enrique Butty (1877–1973); Ramón G. Loyarte (1888–1944) was Professor of Physics at the University of Buenos Aires; Coriolano Alberini (1886–1960) was Professor of Philosophy and dean of the Faculty of Humanities at the University of Buenos Aires. At the time, the university did not yet have a separate Faculty of Engineering. Einstein was referring to Eduardo Huergo, the dean of the Faculty of Exact, Physical, and Natural Sciences, who was an engineer. Other members of the travel party were Ángel Gallardo and Nirenstein and his wife, Magdalena Nirenstein-Holmberg Jorge (1876–1946). Córdoba is located 700 kilometers northeast of Buenos Aires. Einstein was greeted at the Córdoba railway station by academics, government officials, and representatives of Jewish institutions that included the Herzl Society. During his brief visit to the city, Einstein lodged at the Plaza Hotel (see *La Prensa* 13 April 1923, and *Ortiz 1995*, pp. 105–106).

During his visit to Córdoba, Einstein also met with Georg F. Nicolai (see Text 18 in the Additional Texts section of this volume).

55. The banquet was held at the Plaza Hotel and attended by Einstein and his entourage; the rector of the University of Córdoba, Léon S. Morra; faculty members; and the province's governor (*La Prensa*, 13 April 1923). The Sierras de Córdoba mountain range lies to the west of Córdoba.
56. The reception for Einstein was hosted by the rector of the university. The newly elected governor of the province of Córdoba was Ramón J. Cárcano (1860–1946). In Córdoba, Einstein also toured Lake San Roque and lunched at the Edén Hotel de la Falda (see *La Prensa*, 13 April 1923, and *Gangui and Ortiz 2005*, p. 25).
57. Coriolano Alberini.
58. The Córdoba cathedral, the Nuestra Señora de la Asunción, was initially built in the late 16th century and rebuilt in the early 18th century.
59. Einstein does not mention the sixth in his lecture series, which he delivered on 15 April.
60. Einstein was greeted at the offices of the Federación Sionista Argentina by Isaac Nissensohn and Natán Gesang. For the Statement on Nationalism and Zionism he made at the federation, see Text 19 in the Additional Texts section of this volume (see *La Prensa* and *La Época*, 17 April 1925, and *Ortiz 1995*, p. 111).
61. The special session of the Academia Nacional de Ciencias Exactas, Físicas y Naturales was hosted by the president, Eduardo L. Holmberg (see *La Prensa*, 17 April 1925).
62. On the occasion of this lecture, Einstein was presented with a diploma of honorary membership of the Faculty of Exact, Physical, and Natural Sciences by Ramon G. Loyarte (see *La Prensa*, 18 April 1925).

In attendance at the reception were Ángel Gallardo, Minister of Education Sagarno, Minister of Agriculture Tomás Le Breton, positivist philosopher José Ingenieros, civil engineer Nicolás Besio Moreno, musician Carlos López Buchardo, officials of the German Embassy, the president of the German cultural institution Ricardo Seeber, and Nirenstein. For a full list of the invitees, see *Die Deutsche La Plata Zeitung*, 18 April 1925.

63. "Panther cat" refers to Else Jerusalem. "Broges" is Yiddish for "peeved."

The local Jewish community held a reception in Einstein's honor at the Teatro Capitol, a newly restored film theater. Einstein was introduced by Jacobo Saslavsky, president of the Asociación Hebraica, and then delivered a speech. After the speech, a reception was held at the headquarters of the Asociación Hebraica. Einstein was awarded an honorary membership (see *La Prensa*, 19 April 1925).

64. The country *estancia* Llavallol. Jacobo Saslavsky. Robert Koch.

    The dinner was hosted by the local Zionist dignitaries and took place at the Hotel Savoy. Natán Gesang (1888–1944) was a senior member of the Argentinian Zionist Federation and designated as the Zionist movement's coordinator in regard to Einstein's tour. Ben-Zion Mossinson. During his visit, there were contradictory statements on Einstein's alleged identity as a Zionist in the general and Jewish Argentinian press (see *La Prensa*, 19 and 20 April 1925; *Die Presse*, 24, 25 and 27 March 1925; *Crítica*, 26 March 1925; and *Ortiz 1995*, pp. 109–110).
65. On 20 April, Einstein also visited Professor of Applied Mechanics Jorge Duclout, who had been instrumental in inviting him to Argentina and was convalescing (see *Ortiz 1995*, p. 107).
66. The luncheon, held at the Rowing Club Tigre, hosted by dean Eduardo Huergo, was attended by many academics and members of Einstein's reception committee (see *La Prensa*, 21 April 1925).
67. Berta Wassermann-Bornberg.
68. Mauricio Nirenstein.
69. Possibly a reference to Nirenstein's role in assuaging the unease among the wealthy Jewish benefactors of his tour caused by Einstein's political statements at the outset of his visit (see *Gangui and Ortiz 2008*, p. 440).
70. Spanish for sage or scholar.
71. Else Jerusalem.
72. Ángel Gallardo (1867–1934) was the Argentinian minister of foreign affairs. Rector José Arce.

    In the evening, a banquet in Einstein's honor was hosted by the Centro de Estudiantes de Ingeniería in the dining hall of the local branch of the YMCA. Einstein was welcomed by the president of the Centro, Señor Malvicino, and greeted by thunderous applause. The banquet was attended by Eduardo Huergo, Julio Rey Pastor, Julio R. Castiñeiras, Ramón G. Loyarte, Enrique Butty, and a large number of students. According to one report, at one stage during the meal, Einstein participated briefly in throwing bread rolls between the banquet tables. Following the meal, which was accompanied by a Creole orchestra, the guitar teacher Juan Mas played various melancholy and national airs. After a long delay in locating a violin, Einstein played pieces by Schumann, Mozart, and Beethoven (see *La Prensa*, 19 and 23 April 1925; *La Razón*, 24 April 1925; and *Ortiz 1995*, p. 115).
73. Victor Scharf (1872–1943) was an Austrian portraitist, for whom Einstein sat during his visit to Buenos Aires (see Victor Scharf to Einstein, 20 November 1927 [AEA, 48 378]).

According to press reports, Einstein invited Lugones and the physicists Teófilo Isnardi and Ramón G. Loyarte to lunch on the day of his departure. He also visited the Sephardi synagogue, where he was greeted by numerous families of the congregation. At the request of Rabbi Israel Ehrlich, Einstein agreed that an institute for the teaching of Talmudic studies and Hebrew language be named after him. Furthermore, in a comprehensive interview with *La Prensa*, Einstein made extensive comments on his visit to the *La Prensa* offices, on Argentinian culture, scientific research, his impressions of Buenos Aires, La Plata, and Córdoba, and on his concept of Zionism (see *La Prensa* and *La Mañana*, 24 April 1925).

Einstein was bid farewell at the port by a large group of high-ranking university officials, professors, representatives of various cultural, scientific, and Jewish institutions, and students. Among them were Eduardo Huergo, Benito Nazar Anchorena, Julio Rey Pastor, and Enrique Butty. President Anchorena presented Einstein with a diploma recognizing him as an honorary member of the University of La Plata (see *La Prensa*, 24 April 1925).

74. Einstein arrived in Montevideo on board the S.S. *Ciudad de Buenos Aires*, which belonged to the Compañia Argentina de Navegación. He was greeted by, among others, Américo Sampognaro, on behalf of the president of the Republic; Agustín Musso, on behalf of the university president; and Carlos M. Maggiolo (1881–1935), dean of the College of Engineering. A delegation from the local Jewish community and students also welcomed him (see *El Día*, 24 April 1925). The city of Montevideo offered to put Einstein up at the Parque Hotel, yet he had already made arrangements to stay with the family of Naum Rossenblatt, a Russian–Jewish chemist. He was driven to the residence of the Rossenblatt family at Avenida 18 de Julio, the most important thoroughfare in Montevideo, accompanied by Maggiolo. The German ambassador was Arthur Schmidt-Elskop (1875–1952). Einstein and Rossenblatt went for a stroll on the Avenida 18 de Julio, where they happened to meet Carlos Vaz Ferreira (1872–1958), Associate Professor of Philosophy at the University of the Republic, with whom they arranged to meet later that day at Rossenblatt's home. "Ras. Fereider" is most likely an erroneous reference to Vaz Ferreira.

    Einstein was also visited by Teófilo D. Piñeyro, a representative of the Ateneo de Montevideo, a prestigious cultural institution.

    The heads of the Center for Engineering and Land Surveying Students and the Uruguayan Polytechnic Association recommended that their members welcome Einstein at the port. Authorities at the College of Engineering permitted their faculty and students to be absent from classes for this occasion (see *La Prensa* and *La Mañana*, 24 April 1925; *El Día*, 25 April 1925; and *Ortiz and Otero 2001*, pp. 1–2).

75. Esther Rossenblatt-Filevich. José (1896–1953), Octavio, and Gregorio Rossenblatt. José was the first Jewish doctor to graduate from the University of Buenos Aires.

76. Amadeo Geille Castro (1890–?) was Assistant Professor of Rational Mechanics at the University of the Republic in Montevideo. The College of Engineering had designated Geille Castro to serve as Einstein's personal secretary during his visit. He

was assisted in this task by students Ricardo Müller and Ezequiel Sánchez González (see *Ortiz and Otero 2001*, p. 6).

77. In the morning of 25 April, Einstein briefly met with a delegation from the local Jewish community. He then left on a tour of the city by car in the company of Geille Castro, Müller, and Sánchez González, which included a public school housed in the Castro country villa.

    The first lecture was delivered in French in the overcrowded Public Assembly Hall of the University of the Republic at 5:30 p.m. The audience numbered approximately 2,000 people and mainly comprised professors and students. The lecture series was entitled "General Foundations of the Theory of Relativity." Einstein was introduced by the rector of the university, Elías Regules. The engineer Federico García Martínez gave an outline on the theory of relativity. In his lecture, Einstein presented a critique of Newtonian mechanics, dealt with the experiments of Foucault and of Michelson and Morley, outlined the path from special to general relativity, discussed the finiteness of space and of four-dimensional space, the curvature of light rays, and the relativity of time. The reception at the university was hosted by rector Regules and attended by numerous professors and students (*El País*, 23 April 1925; *La Prensa*, 24 April 1925; *El Día*, 24 and 25 April 1925; *La Tribuna Popular*, 26 April 1925; and *La Razón*, 27 April 1925).

78. On 26 April, Einstein held a press conference for three journalists at the Rossenblatt residence, where he elaborated on his literary and musical preferences, his impressions of Montevideo, and the intellectual atmosphere and the state of scientific research in Buenos Aires and Montevideo (see *El País*, 27 April 1925). Montevideo did not have a mayor in 1925, as municipalities had been abolished by the Uruguayan Constitution of 1918. Einstein was referring to the president of the Administrative Council of Montevideo, Luis P. Ponce (1877–1928). *Lohengrin* was performed at the Teatro Solís (see *El Día*, 26 April 1925).

79. The origins of the modern welfare state in Uruguay were established during the period 1904–1916, especially during the presidencies of José Batlle y Ordóñez. The liberal Colorado Party, which was also in power at the time of Einstein's visit, had introduced progressive social legislation. The separation of church and state had been introduced in 1919 with the new Uruguayan constitution (see *Segura-Ubiergo 2007*, p. 58, and *Lynch 2012*, p. 198).

80. The president of the Senate was Juan Antonio Buero (1888/9?–?). The marble factory was the Compañia de Materiales para Construcción in the Bella Vista neighborhood. The new government building was the Legislative Palace designed by Vittorio Meano and Gaetano Moretti (see *El Día*, 28 April 1925).

81. The audience with the president of the Republic, José Serrato (1868–1960), took place at 3 p.m.

    The minister of justice and public instruction was José Cerruti. The Swiss consul in Montevideo was possibly Maximo (Max) Guyer.

    The second lecture was also held at the Public Assembly Hall of the university at 5:30 p.m. The audience was even larger than at the first lecture. Einstein first continued with his exposition of the special theory of relativity, dealing with the constancy of the speed of light, the validity of natural laws for all inertial systems, and the Lorentz transformation. He then proceeded to elaborate on the general theory of relativity, discussing gravitational fields and relative acceleration (see *La Mañana*, 28 April 1925).

82. The Federation of German Associations had decided "unanimously" to greet Einstein by means of a reception committee and to hold a reception in his honor at the German Club (see German Embassy, Montevideo, to the German Foreign Ministry, Berlin, 4 June 1925 [GyBPAAA, R 64678]).

    The banquet in Einstein's honor was held by the local Jewish community at 8 p.m. at the Hotel del Prado. Fridtjof Nansen (1861–1930) was a Norwegian explorer, scientist, humanitarian, and the League of Nations' High Commissioner for Refugees. Jakob Wolf Latzki-Bertholdi (1881–1940) was a journalist and the representative in South America of Emigdirect, the United Jewish Aid Society of Europe. The Argentinian Foreign Minister Ángel Gallardo (see *El Día*, 29 April 1925).

83. Einstein arrived at the reception at the College of Engineering at 10 a.m. He was accompanied by members of the college's council, Professors of Engineering Carlos Berta and Bernardo Larrayoz, and Geille Castro and his assistants. Álvarez Cortés, the minister for public works, was also present. Einstein was greeted by the dean of the College, Donato Gaminara, and by a large group of students and professors. He toured the college's laboratories and library and was presented with a diploma and a gold plaque. The diploma recognized Einstein as an honorary member of the Association of Engineering and Landsurveying Students (see *El País*, 29 April 1925, and *Ingeniería*, Vol. 17 [1925], 4). In the early evening, Einstein visited the National Senate, where he was greeted by its president, Juan Antonio Buero.

    In his third and final lecture, Einstein again dealt with the general theory of relativity, including the role of Gaussian and Riemannian mathematics in developing his theory and the experimental proofs of his theory (see *Ortiz and Otero 2001*, p. 19).

    The reception was hosted by the German ambassador, Arthur Schmidt-Elskop. Among the attendees were Luis Alberto de Herrera, president-elect of the National Council, and Juan Carlos Bianco, Minister of Foreign Relations. Reports in the press listed the names of Uruguayan politicians and scholars but did not mention prominent Germans (see *El Bien Público*, 30 April 1925). The ambassador expressed his pleasure that Einstein was referred to as the *sabio alemán* (German scholar) in the Uruguay press (see German Embassy, Montevideo, to the German Foreign Ministry, Berlin, 4 June 1925 [GyBPAAA, R 64678]).

84. *The Pilgrim* was produced and directed by Charlie Chaplin in 1923. Max Glücksmann (1875–1946) was an Austrian-born Jewish pioneer of the Argentinian music and film industries in Buenos Aires.

85. At the reception of the Polytechnic Association, its president, Victor V. Sudriers, informed Einstein that he was to be elected an honorary member of the association.

    The banquet was hosted by the University of the Republic and held at the Hotel del Prado. Among the attendees were President José Serrato, ministers of state, members of the High Court, the Senate, and the Chamber of Deputies, the German ambassador, and university professors. Einstein was presented with a diploma [AEA, 65 036] that appointed him as an honorary professor of the university. The banquet was followed by musical performances in the hotel's ballroom (see *La Mañana*, 29 April 1925; *Ingeniería*, Vol. 1 [1925], *La Razón*, 29 April 1925; *El Día*, 2 May 1925; and *Ortiz and Otero 2001*, p. 15).

"Die Wacht am Rhein" (The Watch on the Rhine) is a militaristic German anthem written in the mid-19th century at a time of French–German enmity. During the German Empire, it served as an unofficial second anthem. The German national anthem in the Weimar Republic was the "Deutschlandlied," which had been adopted in 1922.

86. The S.S. *Valdivia* belonged to the Société Générale des Transports Maritimes à Vapeur.
87. Einstein was bid farewell by a large number of professors and students at the dock (see *El Día*, 10 May 1925).
88. In a section of his book titled "Les illusions socialistes," Le Bon argues that socialism is irreconcilable with democracy and that its triumph would lead to universal enslavement. He also believes that combative struggle is necessary for human progress and that pacifism would lead to the extinction of nations (see *Le Bon* 1919, pp. 155–159 and 161–163).
89. See Texts 18 and 21 in the Additional Texts section of this volume for references to an idea Einstein had while in Argentina.
90. Retired general Paul von Hindenburg (1847–1934) had been elected president of Germany on 26 April 1925 (see *Berliner Tageblatt*, 27 April 1925, evening edition).
91. The bay of Rio de Janeiro is studded with 130 islands.
92. Félix d'Herelle, who had discovered bacteriophage in 1917, published a series of articles on phage therapy in 1921–1925, based on his experiments at the Institut Pasteur in Paris. See *Summers 1999*, chap. 8.
93. Einstein was greeted at the port by Rabbi Isaiah Raffalovich, Isidoro Kohn, and Daniel Henninger; Arthur Getúlio das Neves, Professor of Engineering at the Polytechnic School and deputy president of the Engineering Club; Aloísio de Castro; and several students. He was taken directly to the upscale Hotel Glória on the waterfront to rest for a while. In discussions with his reception committee, it was decided that Einstein would give two lectures on the theory of relativity—one at the Engineering Club and one at the Polytechnic School, tour a number of scientific institutions, and reserve one evening for a reception with the Jewish community. His visit would also include an audience with the president of Brazil, a rail journey to the city of Itatiaia in the mountains, and ample time to rest (see *O Jornal*, 5 May 1925, and *Tolmasquim 2003*, pp. 127–130).
94. Isidoro Kohn. Einstein uses the term "Gschaftlhuber," Bavarian or Austrian slang for a braggadocio.
95. Irma Kohn. Poldi Wettl. On their walk, Kohn suggested that they purchase a tailcoat for Einstein's audience with the president of the Republic the next day (see *Tolmasquim 2003*, p. 130).

96. The German merchants invited Einstein to a meeting with representatives of the local German community. He also received an honorary doctorate of philosophy from Washington Garcia, dean of the School of Philosophy at the University of Rio de Janeiro (see Washington Garcia to Einstein, 5 May 1925 [*CPAE 2015*, Vol. 14, Abs. 679] and diploma [AEA, 65 038]).

 The group of professors who accompanied Einstein on his outing to Sugarloaf Mountain included Aloísio de Castro and Getúlio das Neves (see *Jornal do Brasil*, 6 May 1925, and *Tolmasquim 2003*, pp. 131 and 239).

97. Antônio da Silva Mello (1886–1973) was Professor of Clinical Medicine at the University of Rio de Janeiro and a reformer of the Brazilian medical system. They walked in the Santa Teresa neighborhood and had lunch at Restaurante do Minho (see *Tolmasquim 2003*, p. 132).

98. The president of the Republic was Artur Bernardes (1875–1955). The delegation to the president was headed by Getúlio das Neves. The meeting was primarily a courtesy visit. Toward the end of the audience, the president had to deal with suppressing a revolt by the younger generation of the Brazilian military. Einstein posed for photographs with the president and various cabinet ministers, including Minister of Justice Afonso Pena Júnior. The Brazilian Ministry of Education was only established in 1931 (*World Bank 2002*, p. 65). The mayor of Rio de Janeiro was Alaor Prata Soares (1882–1964).

 Einstein's first public lecture on relativity at the Engineering Club was attended by the American and Portuguese ambassadors, anti-relativist admiral Gago Coutinho and several generals from the armed forces, representatives of the Ministers of Justice and Agriculture, the mayor of Rio de Janeiro, and prominent engineers and physicians. Many wives and children were also in attendance. Einstein was briefly introduced by Getúlio das Neves. He delivered the lecture in French. According to press reports, some members of the audience were seated so close to Einstein that he was pushed against the blackboard. Following his lecture, Einstein was presented with four bronze medals that commemorated the centenary of the Brazilian republic in 1922 (see *O Imparcial*, *Jornal do Brasil*, and *O Jornal*, 7 May 1925; *Caffarelli 1979*, pp. 1439 and 1441–1442; and *Tolmasquim 2003*, pp. 132–135).

99. During his visit to the National Museum of Brazil, Einstein was greeted by the anthropologist Edgar Roquette-Pinto, since the director of the museum, Arthur Neiva, was in São Paulo at the time. Among the various objects shown to Einstein was the ten-ton Bendegó meteorite. Roquette-Pinto gave Einstein an indigenous bracelet as a gift for his wife. Thereupon, Einstein allegedly asked for another one for Mileva (see *O Paiz*, 12 May 1925, and *Glick 1999*, pp. 109–110).

 The luncheon hosted by Aloísio de Castro was also attended by the physicians Miguel Couto and Silva Mello, Getúlio das Neves and Daniel Henninger; the Russian archeologist Alberto Childe (1870–1950); the feminist author Rosalina Coelho Lisboa (1900–1975); and Assis Chateaubriand (see *Jornal do Brasil*, 9 May 1925).

The Brazilian Academy of Sciences had been established in 1916. The reception was attended by approximately 100 people, including numerous academics, officials from various political institutions and universities, the ambassadors of Germany, Great Britain, and the United States, professors and students. Einstein was first greeted by the deputy president of the academy, Juliano Moreira. The president, Henrique Morize, was present but unable to speak due to medical problems. Moreira presented Einstein with a diploma confirming his status as the first corresponding member of the academy [AEA, 65 037]. The second speaker at the reception was member of the academy Francisco Lafayette Rodrigues Pereira, who presented a brief overview of Einstein's theories (for his speech, see *O Jornal*, 7 May 1925). The last speaker was member of the academy Mario Ramos, who announced that the academy planned to establish an Einstein prize, to be awarded annually for outstanding work in the sciences (see *Jornal do Brasil*, 8 May 1925). Einstein responded by presenting a lecture on the current state of research on the theory of light. He prepared his talk on stationery of the Hotel Glória and gave the manuscript to das Neves.

Following the reception, Einstein drove with a group of academy members to the studios of Rádio Sociedade, where he recorded a short address (see Text 27 in the Additional Texts section of this volume; see also *Jornal do Brasil*, 8 May 1925; *Tolmasquim 2003*, pp. 135–140; and *Tolmasquim and Moreira 2002*, p. 234).

100. The Oswaldo Cruz Institute, devoted to the study of tropical diseases, had been established according to the model of the Pasteur Institute in Paris. Einstein was greeted by Carlos Chagas and Leocádio Chaves, director and deputy director of the institute, respectively. He toured the Oswaldo Cruz Museum, the Museum of Pathological Anatomy, laboratories, and the library. He was also shown the pathogen that causes the tropical illness called Chagas' disease (see *Gazeta das Noticias*, 9 May 1925).

Einstein gave his second public lecture on relativity at the Polytechnic School. This time the audience was more limited, mainly consisting of the faculty and special guests. He was introduced by the school's acting director, José Agostinho dos Reis. In content, the second lecture was a continuation of the first (see *O Paiz* and *O Jornal*, 9 May 1925; *Caffarelli 1979*, p. 1443).

The supper, hosted by the German ambassador, Hubert Knipping (1868–1937), was held at the Club "Germania," established in 1821. The event was attended by the president of the German Chamber of Commerce, various businessmen, industrialists and bankers, and Isidoro Kohn. One newspaper reported that Einstein stated that, similar as in Europe, he had noticed indications of suspicion among the nations of the Americas, but that there was less friction due to a higher degree of tolerance. In a report to Berlin, Knipping informed the German Foreign Ministry that the Brazilian foreign minister had been invited but had decided to send a representative instead (see *O Jornal*, 9 May 1925; Report of the German Embassy Rio de Janeiro, 20 May 1925 [GyBPAAA, R 64678]; *Tolmasquim and Moreira 2002*, p. 230; *Tolmasquim 2003*, pp. 140–143; and *Glick 1999*, p. 110).

101. The National Observatory, established in 1827, had recently moved to a new campus in the neighborhood of São Cristóvão. Einstein was accompanied by Ignácio do Amaral, Professor at the Polytechnic School; the civil engineer Alfredo Lisboa; and Isidoro Kohn. They were welcomed by the director of the observatory, Henrique Morize. Einstein met with local astronomers who had participated in the British eclipse expedition to Sobral, including Lélio Gama, Allyrio de Mattos, Domingos Costa, and Morize himself. At the observatory, the transmission of a signal that recorded standard time and a Milne-Shaw seismograph, the standard recording instrument at the time, were demonstrated to Einstein.

The two brothers were Álvaro and Miguel Ozório de Almeida.

Einstein had dinner at the home of Isidoro Kohn. The reception of the local Jewish community was held at the Automobile Club of Brazil. Einstein may have been thinking of the Jockey Club in Buenos Aires, where a reception had been held in his honor (see note 48 above). The event was attended by over 2,000 people. Isaiah Raffalovich welcomed Einstein in German. Other speakers were Jóse David Perez for the Sephardi community, who addressed Einstein in Portuguese, and Eduardo Horovitz for the Ashkenazi community, who addressed him in Yiddish. In his response, Einstein thanked his hosts and spoke of the great significance of solidarity among Jews and of their importance in the Zionist colonization efforts in Palestine (for the speeches of Raffalovich and Perez and a paraphrase of Einstein's response, see *O Jornal*, 9 May 1925; *Jornal do Brasil*, 10 May 1925; *O Paiz*, 15 May 1925; *Dos Idische Vochenblat*, 22 May 1925; and *Tolmasquim 2003*, pp. 147 and 219–221; see also *A Noite*, 11 May 1925, and *Glick 1999*, pp. 105 and 110–111).

Robert Koch, who lived in Buenos Aires.

102. Originally, a train trip to Itatiaia had been planned (see note 93 above). However, it was deemed that such a trip would be too long, and instead a tour by car in the environs of Rio de Janeiro was organized. Einstein drove with Rabbi Raffalovich, Isidoro and Irma Kohn, and Poldi Wettl. A second car transported Getúlio das Neves, Antonio Pacheco Leão, Ignácio do Amaral, Mário de Souza, and Roberto Marinho de Azevedo. They drove along the coast through Copacabana, down Avenida Niemeyer, and up the Dois Irmãos Hill. After lunch, they drove through the neighborhood of São Conrado, toured Tijuca Forest, and took the funicular up to Corcovado Mountain.

In the evening, Einstein visited the Zionist Center and the Scholem Aleichem Library. At the former, he was welcomed by Jacobo Schneider, president of the Zionist Federation, and Moisés Koslovski, president of the Zionist Center. In his response, Einstein expressed his satisfaction with the intense Zionist activity of the local Jewish community. At the latter, he was greeted by the president of the library, Lewis Feingold, was given a leather-bound edition of Scholem Aleichem's works, and presented the library with an autographed photograph of himself (see *Tolmasquim 2003*, pp. 148–150).

103. The National Hospital for the Insane was originally established in 1852 as the first psychiatric hospital in Latin America. Einstein was welcomed by its director, Juliano

Moreira (1873–1932), a proponent of more humane psychiatric treatment methods. At the hospital, Einstein requested to see a patient with a case of paranoia.

Moreira's wife was Augusta Moreira-Peick (1876–1950). The lunch, at which vatapá—an Afro-Brazilian shrimp dish—was served, was also attended by Isidoro Kohn and Edgar Roquette-Pinto. In the afternoon, Einstein visited the Brazilian Press Association, where he was presented with a gift of precious gems and a check, the monies for which had been raised in a campaign initiated by *O Jornal* (See *O Jornal* to Einstein, 11 May 1925 [*CPAE 2015*, Vol. 14, Abs. 685]). General Cândido Rondon (1865–1958) led an expedition to install telegraph lines on the frontier with Bolivia and Paraguay, and was best known for his advocacy of the indigenous tribes he encountered on his expedition. At the end of his visit, Einstein signed the association's visitors' book, expressing his appreciation for the kindness shown him during his visit and extending a warm embrace to all [*CPAE 2015*, Vol. 14, Abs. 682].

The dinner hosted by the German ambassador was held at the Hotel Glória. Among the attendees were Juliano Moreira, Aloísio de Castro, Ignácio do Amaral, Mário de Souza, Daniel Henninger, and Isidoro Kohn. The German ambassador wished Einstein a safe voyage back to Germany (see *A Noite*, 11 May 1925; *O Jornal* and *O Paiz*, 12 May 1925: *Glick 1999*, p. 106: and *Tolmasquim 2003*, pp. 150–153).

Additional Texts

1. TLS. [AEA, 45 084]. Published in *CPAE 2015*, Vol. 14, Doc. 138, pp. 224–225. Written on personal letterhead, and addressed "Herrn Prof. Dr. Albert Einstein Berlin W. 30 Haberlandstrasse 5." Straus (1876–1956) was director of Carl Lindström Actien-Gesellschaft.
2. TLS. [AEA, 43 089]. Published in *CPAE 2015*, Vol. 14, Doc. 193, pp. 310–311. Original in French. Written on letterhead "Asociación Hebraica, Suipacha 1008, Buenos Aires." A card is attached: "I. Starkmeth, Ayacucho 800, Buenos Aires T. Juncal 1740."
3. Max Straus. Elsa Einstein.
4. This sentence is underlined in red pencil.
5. Elsa Einstein had apparently informed Straus that Einstein would only accept the invitation if it were extended by an academic institution, rather than a private one (see *Ortiz 1995*, pp. 82–83 and 92, and *Tolmasquim and Moreira 2002*, p. 231).
6. Three clippings were enclosed with this document. The other universities were the National Universities of Córdoba, La Plata, the Littoral, and Tucumán (see [AEA, 43 090], [AEA, 43 091], and [AEA, 43 092]). The university council of the University of Buenos Aires had approved the decision to issue a joint invitation from all Argentinian universities at a meeting on 21 December 1923 (see *Ortiz 1995*, pp. 83 and 92).
7. The clippings cited a donation from the Asociación Hebraica of 4,660 Argentinian pesos, equivalent to $1,500 at the time. In comparison, Julio Rey Pastor, a prominent faculty member at the University of Buenos Aires, was earning 1,500 pesos a month.

Einstein's honorarium and travel fares almost equaled the annual salary of a distinguished visiting professor. Funding amounting to 1,500 pesos was also received from the Argentinian–German Cultural Institution (see *Ortiz 1995*, pp. 83–84).

8. For the invitations for a series of lectures from José Arce, the rector of the National University of Buenos Aires, and from Mauricio Nirenstein, see José Arce and Mauricio Nirenstein to Einstein, 31 December 1923 and 7 January 1924 [*CPAE 2015*, Vol. 14, Abs. 238 and 250]. The interim Argentinian chargé d'affaires Pedro Guesalaga forwarded *CPAE 2015*, Abs. 238 to Einstein (see Pedro Guesalaga to Einstein, 14 February 1924 [*CPAE 2015*, Vol. 14, Abs. 303]).
9. This was equivalent to approximately $320 at the time. The university's College of Engineering had asked Uruguay's National Administrative Council for 1,000 pesos to cover the costs of Einstein's visit to Montevideo. However, the council agreed to contribute only 500 pesos. This amount was comparable to the average sum a public speaker would earn (*La Mañana*, 30 April 1925, and *El Sol*, 9 May 1925).
10. Baron Maurice (Moritz) de Hirsch (1831–1896) was a German–Jewish philanthropist and the founder of ICA, the Jewish Colonization Association. ICA had purchased land in the provinces of Buenos Aires, Santa Fe, and Entre-Rios. The director of ICA in Buenos Aires was Isaac Starkmeth (1868–1938).
11. Mauricio Nirenstein.
12. ALSX. [AEA, 88 016]. Published in *CPAE 2015*, Vol. 14, Doc. 222, pp. 347–348. The letter is addressed: "An die Asociation Hebraica, Buenos Aires."
13. Einstein's most significant recent absences from Berlin were his extensive trip to the Far East from October 1922 to March 1923 and his prolonged sojourn in the Netherlands from early November to late December 1923 (see *Einstein 2018*, and *CPAE 2015* Vol. 14, Chronology, entries for 7 or 8 November and 23 December 1923).
14. ALSX. [AEA, 10 089]. Published in *CPAE 2015*, Vol. 14, Doc. 285, pp. 442–444. Ehrenfest (1880–1933) was Professor of Theoretical Physics at the University of Leyden.
15. Ehrenfest had suggested this in Paul Ehrenfest to Einstein, 25 May 1924 [*CPAE 2015*, Vol. 14, Doc. 255].
16. For the invitation, see Text 2 in the Additional Texts section of this volume.
17. ALSX. [AEA, 29 401]. Published in *CPAE 2015*, Vol. 14, Doc. 389, p. 598. Maja Winteler-Einstein (1881–1951) was Einstein's sister. Paul Winteler (1882–1952) was Maja's husband and a retired employee of the Swiss Federal Railroad.
18. Dated on the assumption that this is a reply to Maja Winteler-Einstein to Einstein, 4 December 1924 [*CPAE 2015*, Vol. 14, Doc. 388] and Paul Winteler to Einstein, 4 December 1924 [*CPAE 2015*, Vol. 14, Abs. 563].
19. TLS. [AEA, 44 010]. Abstract published in *CPAE 2015*, Vol. 14, Abs. 640, p. 898.
20. ALS (Dora Kubierschky, Aschau, Germany). [AEA, 84 235]. Published in *CPAE 2015*, Vol. 14, Doc. 439, p. 672. Anschütz-Kaempfe (1872–1931) was the founder and owner of a German company in Kiel which produced nautical and aerial navigational instruments, and a benefactor of Einstein's.

21. AKS. [AEA, 143 179]. Published in *CPAE 2015*, Vol. 14, Doc. 456, p. 709. The card is addressed "Frau Elsa Einstein Haberlandstrasse 5 Berlin" and postmarked "Hamburg 5 3 25 12–1N[achmittags]."
22. Probably Emily Robinow-Kukla and Wilhelm Roloff.
23. Carl Melchior.
24. The S.S. *Cap Polonio*.
25. Hans Mühsam (1876–1957) was a staff physician at the Jewish Hospital in Berlin. See next document, and note 35.
26. Margot Einstein had originally planned to accompany Einstein on his trip but had apparently fallen ill shortly before his departure.
27. See next document, and note 34.
28. ALS. [AEA, 143 180]. Published in *CPAE 2015*, Vol. 14, Doc. 457, pp. 709–711. Written on letterhead "Hamburg-Südamerikanische Dampfschifffahrts-Gesellschaft. 'Cap Polonio.'"
29. Wilhelm Roloff.
30. Alexander Bärwald (1877–1930) was a German–Jewish architect.
31. Possibly Louis Lewin (1850–1929), Professor of Pharmacology at the Technical University in Berlin.
32. Carl Jesinghaus.
33. Rudolf Kayser (1889–1964) was editor in chief of the German literary journal *Die Neue Rundschau* and Einstein's son-in-law.
34. Enrique Gaviola (1900–1989) was a student of physics at the University of Berlin.
35. The letter and the cash may be related to the dramatic termination of the romantic affair between Einstein and his former secretary Betty Neumann. Hans Mühsam was a cousin of Neumann's (see the section "Einstein's Acceptance of the Invitations" in the Historical Introduction).
36. Presumably a reference to Einstein's previous abdominal ailments which led to his need to adhere to a strict diet (see Hans Mühsam to Einstein, 9 December 1923 [*CPAE 2015*, Vol. 14, Doc. 174]).
37. Elsa's parents, Rudolf (1843–1926) and Fanny (1852–1926) Einstein. "Rudilse" was a nickname for Rudolf Kayser and Ilse Kayser-Einstein.
38. Tgm. [AEA, 143 182]. Published in *CPAE 2015*, Vol. 14, Doc. 461, p. 714. Typed on the telegram form of "Deutsche Betriebsgesellschaft für drahtlose Telegrafie m.b.H., Berlin SW. 61."
39. Einstein had turned 46 on 14 March.
40. ALS. [AEA, 143 183]. Published in *CPAE 2015*, Vol. 14, Doc. 462, p. 715. Written on letterhead "Hamburg-Südamerikanische Dampfschifffahrts-Gesellschaft 'Cap Polonio.'"
41. Else Jerusalem and Carl Jesinghaus.
42. See text of diary, this volume, entry for 19 March 1925.
43. While traversing the Red Sea on 15 October 1923 on his voyage to the Far East, Einstein had described the heat there as "[g]reenhouse temperature" (see *Einstein 2018*, p. 95).
44. Margot Einstein.

45. Einstein had toured Lisbon on 11 March, and sailed past the Teide Peak volcano on Tenerife Island on 12 March (see text of diary, this volume, entries for 11 and 12 March 1925).
46. Max Planck (1858–1947) was Professor of Physics at the University of Berlin.
47. ALS. [AEA, 143 184]. Published in *CPAE 2015*, Vol. 14, Doc. 464, p. 723.
48. Dated by reference to Einstein's arrival in Buenos Aires the day before (see text of diary, this volume, entry for 24 March 1925).
49. Bruno Wassermann. The name of the street is actually Zabala.
50. Rudolf and Fanny Einstein.
51. Antônio da Silva Mello. Rudolf Ehrmann (1879–1963) was Professor of Internal Medicine at the University of Berlin.
52. Moritz Katzenstein. Probably Louis Lewin.
53. In a question and answer session with local journalists in Buenos Aires, Einstein was asked about his position on Zionism. His statement was published in an article entitled "Einstein y sus verdugos" (Einstein and His Executioners) and in *Mundo Israelita*, 28 March 1925, p. 2. Published in *CPAE 2015*, Vol. 14, Appendix E, p. 914.
54. The Hebrew University was to be inaugurated on 1 April 1925.
55. ALS. [AEA, 143 185]. Published in *CPAE 2015*, Vol. 14, Doc. 471, pp. 741–742. Written on letterhead "Bruno John Wassermann 579 Azopardo Buenos Aires."
56. See text of diary, this volume, entry for 1 April 1925.
57. Bruno Wassermann and Berta Wassermann-Bornberg, Einstein's hosts in Buenos Aires.
58. Einstein planned to hold a lecture at the National University of Córdoba.
59. Robert Koch, Louis Louis-Dreyfus (see text of diary, this volume, notes 42 and 49).
60. Georg F. Nicolai (1874–1964) was Professor of Physiology at the National University of Córdoba. They had cosigned the pacifist "Manifesto to the Europeans" in 1914 (see *CPAE 1996*, Vol. 6, Doc. 8, mid-October 1914). See also note 79 below.
61. Lisa Hirsch-Gottschalk.
62. Alfredo Hirsch.
63. Ángel Gallardo.
64. This speech was delivered at a celebration of the inauguration of the Hebrew University of Jerusalem at the Teatro Coliseo in Buenos Aires on 6 April 1925 and published under the title "La colectividad israelita celebró la creación de la Universidad de Jerusalén" (The Jewish Community Celebrates the Establishment of the University of Jerusalem) in *La Prensa*, 7 April 1925. Published in *CPAE 2015*, Vol. 14, Appendix H, pp. 972–973.
65. For Einstein's description of the event and the press coverage, see text of diary, this volume, entry for 6 April 1925, and note 51.
66. Theodor Herzl (1860–1904) was the founder of modern political Zionism. Chaim Weizmann (1874–1952) was the president of the World Zionist Organization.
67. AKS. [AEA, 122 774]. Published in *CPAE 2015*, Vol. 14, Doc. 472, p. 742. The card is addressed "Herrn und Frau Dr. Kayser (Schlemihlim) Kaiserallee 32 Berlin-Schöneberg

Germany" and postmarked "Buenos Aires Abr 13 1925 20*." There is a photo of a bathing resort in Buenos Aires on the verso. "Schlemihlim" is a variant of *shlemiehls*.

68. At the Wassermann's country estate in Llavallol (see text of diary, this volume, entry for 8–10 April 1925).
69. Elsa, her daughters, and her parents all lived at Haberlandstraße 5 in Berlin.
70. AKS. [AEA, 143 283]. The card is addressed "Frl. Margot Einstein Haberlandstr. 5 Berlin (Germany)" and postmarked "Buenos Aires Abr 13 1925 20*."
71. At Llavallol.
72. ALS. [AEA, 143 186]. Published in *CPAE 2015*, Vol. 14, Doc. 474, pp. 743–744. Written on letterhead "Bruno John Wassermann, 579 Azopardo, Buenos Aires."
73. See text of diary, this volume.
74. Bruno Wassermann and Berta Wassermann-Bornberg.
75. He was actually due there on 4 May.
76. The S.S. *Cap Norte*, which belonged to the Hamburg South American line.
77. On Einstein's reception by the German community in Buenos Aires, see text of diary, this volume, entry for 17 April 1925.
78. Carl Gneist.
79. Georg F. Nicolai, who had a child out of wedlock with Elly von Schneider (1887–1967), a friend of Einstein's step-daughter Ilse Kayser-Einstein. The child was Arne von Schneider-Glend (1922–2001).
80. On the luncheon, see text of diary, entry for 2 April 1925, this volume, and note 47.
81. The idea had to do with the relation between electricity and gravitation (see Lecture Notes, ca. 18 April 1925 [*CPAE 2015*, Vol. 14, Doc. 475]). On 2 May, however, Einstein noted in his travel diary that all solutions he had entertained while in Argentina had turned out to be useless.

    The earlier idea Einstein refers to here is mentioned in the entry of 6 January 1923 in the travel diary of his trip to Japan (*Einstein 2018*, p. 197).
82. Rudolf Ehrmann was visiting Palestine to attend the opening of the Hebrew University (see Einstein to Herbert L. Samuel, 2 March 1925 [*CPAE 2015*, Vol. 14, Doc. 454]).
83. The Einstein-Balfour Physics and Mathematical Institute was inaugurated at the Hebrew University on 2 April (see Chaim Weizmann to Einstein, 2 April 1925 [*CPAE 2015*, Vol. 14, Doc. 469]). A donation of $100,000 for the establishment of the institute was received from the New York businessman Philip Wattenberg (see *New York Times*, 6 April 1925).
84. The manuscript of *Einstein 1916* [*CPAE 1996*, Vol. 6, Doc. 30]. The issue of the disposition of the manuscript had been deliberated since 1922 (see Paul Oppenheim to Einstein, 9 April 1922 [*CPAE 2012*, Vol. 13, Doc. 138]).
85. For details of Einstein's honoraria for his lecture tour in Argentina, see Text 2 in the Additional Texts section of this volume.
86. This statement was made during Einstein's visit to the offices of the Zionist Federation in Buenos Aires on 16 April 1925 and published in an article entitled "Visita del profesor Einstein a la Federación Sionista" (Visit of Professor Einstein at the

Zionist Federation) in *Mundo Israelita* on 18 April 1925, p 1. Published in *CPAE 2015*, Vol. 14, Appendix I, p. 974.

87. ALS. [AEA, 143 187]. Published in *CPAE 2015*, Vol. 14, Doc. 477, pp. 746–477. Written on letterhead of "Bruno John Wassermann 579 Azopardo Buenos Aires."
88. Bruno Wassermann and Berta Wassermann-Bornberg.
89. Presumably 20,000 Reichsmarks, which equaled $4,656.58 according to the exchange rate at the time (see *Vossische Zeitung*, 25 April 1925). Einstein had been offered $4,000 as a joint honorarium from five Argentinian universities, and 1,000 pesos from the University of Montevideo. In addition, he had been offered $700 from *La Prensa* for a series of five articles (see Text 2 in the Additional Texts section in this volume and Carl D. Groat to Einstein, 27 November 1924 [*CPAE 2015*, Vol. 14, Abs. 539]).
90. Rudolf Kayser and Ilse Kayser-Einstein. Margot Einstein.
91. In Buenos Aires, Einstein had attended a festive celebration on the occasion of the opening of the Hebrew University, had visited the offices of the local Zionist executive, had given a lecture on Zionism, and had attended a large Zionist reception (see text of the diary, this volume, entries for 6, 16, 18, and 19 April 1925).
92. See text of the diary, this volume, note 73.
93. A week earlier, Einstein had expressed his opposition to the donation of the manuscript of *Einstein 1916*. In the meantime, Elsa had arranged for its donation to the Hebrew University on the occasion of its official opening (see Text 18 in the Additional Texts section of this volume, and Leo Kohn to Einstein, 19 March 1925 [*CPAE 2015*, Vol. 14, Abs. 665]).
94. Rudolf and Fanny Einstein.
95. ALSX. [AEA, 75 631]. Published in *CPAE 2015*, Vol. 14, Doc. 476, p. 746. Written on letterhead of "Bruno John Wassermann 579 Azopardo Buenos Aires."
96. Mileva Einstein-Marić.
97. See Text 18 in the Additional Texts section in this volume and note 81 for more on this idea.
98. To stay at his bachelor's apartment at the factory of Hermann Anschütz-Kaempfe.
99. Possibly due to the stomach ailments he had suffered the previous year (see Heinrich Zangger to Einstein, after 9 May 1924 [*CPAE 2015*, Vol. 14, Doc. 243]).
100. ALS. [AEA, 143 188]. Published in *CPAE 2015*, Vol. 14, Doc. 478, pp. 747–748. Written on letterhead of "Naum Rossenblatt Montevideo."
101. Naum Rossenblatt.
102. The S.S. *Valdivia*.
103. On the lecture, see text of diary, this volume, note 81. The minister of foreign affairs was Juan Carlos Bianco. The president of the Republic was José Serrato. The president of the Senate was Juan Antonio Buero.
104. For the reception by the German community in Montevideo, see text of diary, this volume, entry for 28 April 1925. For the reception in Buenos Aires, see text of diary, this volume, entry for 17 April 1925, and Text 18 in the Additional Texts section of this volume.

105. The president of the Administrative Council of Montevideo, Luis P. Ponce. See text of diary, this volume, entry for 26 April 1925.
106. Margot Einstein.
107. Published in Spanish in *La Prensa*, 28 April 1925, p. 10, and in *CPAE 2015*, Vol. 14, Doc. 479, pp. 749–752.
108. See, e.g., Einstein's impressions of Japanese culture, society, and intellectual life in "Musings on My Impressions in Japan" [*CPAE 2012*, Vol. 13, Doc. 391].
109. He was reading Vaz Ferreira's criticism of pragmatism (*Vaz Ferreira 1914*) at the time of writing this article (see the next document).
110. A quote from Goethe's poem "Zahme Xenien IX" (see *Goethe 1952*, p. 367).
111. ALS (A. Rossani, Instituto de Profesores "Artigas," Montevideo). [AEA, 88 054]. Published in *CPAE 2015*, Vol. 14, Doc. 480, pp. 753–754. Written on letterhead "Naum Rossenblatt Montevideo." Original in French.
112. Einstein had met Vaz Ferreira a few days earlier (see text of diary, this volume, note 74).
113. *Vaz Ferreira 1914*. Vaz Ferreira's book on pragmatism and its criticism was based on a series of six lectures that had been part of his philosophy course at the University of Montevideo in 1908. The Spanish version of the book had been published in 1909 (see *Vaz Ferreira 1909*).
114. In his critique of pragmatism, Vaz Ferreira writes: "James's [and other pragmatists'] fundamental confusion consisted in pretending to draw practical consequences from that which ought to have been no more than an explanation of truth" (see *Vaz Ferreira 1914*, p. 58).
115. Vaz Ferreira quotes extensively from *James 1897* and *James 1907* and other works, although the references are sometimes inexact or missing.
116. Vaz Ferreira addresses the pragmatists' distinction between "reality" and "truth," e.g., "Reality exists; it is; and we, others, we know it or we do not know it [. . .]. In contrast, according to pragmatism, truth is in the making, it becomes, it is produced, and depends, to a greater or lesser degree, on our facts and even on our beliefs" (see *Vaz Ferreira 1914*, p. 12).
117. AKS. [AEA, 10 104]. Published in *CPAE 2015*, Vol. 14, Doc. 482, p. 764. The card is addressed "Herrn Prof. Dr. P. Ehrenfest Witte-Rozen-Str. Leiden Holland (Europa)" and postmarked "Rio de Janeiro 4. V. 1925." The verso depicts a vista of Rio de Janeiro from Avenida Niemayer.
118. The correct date is 4 May (see the postmark).
119. Einstein had arrived in South America on 21 March 1925 (see text of diary, this volume, entry for 22 March 1925).
120. Ehrenfest's elder daughter Tatiana (1905–1984) had traveled to the Soviet Union in January (see Tatiana Ehrenfest-Afanassjewa to Einstein, 28 November 1924 [*CPAE 2015*, Vol. 14, Abs. 543], and Paul Ehrenfest to Einstein, 8 January 1925 [*CPAE 2015*, Vol. 14, Doc. 415]).
121. AKS. [AEA, 143 189]. Published in *CPAE 2015*, Vol. 14, Doc. 483, p. 764. The card is addressed "Frau Elsa & Frl. Margot Einstein Haberlandstr. 5 Berlin Germany" and

postmarked "Rio de Janeiro [——]." The verso depicts a vista of the Botafogo neighborhood in Rio de Janeiro and includes a note in Einstein's handwriting: "Keep the postcard because it's pretty."

122. From Uruguay on board the S.S. *Valdivia* (see text of diary, this volume, entries for 1 and 4 May 1925).
123. The Hotel Glória.
124. Ilse Kayser-Einstein.
125. Probably Louis Lewin.
126. Possibly Fritz Haber (1868–1934), director of the Kaiser Wilhelm Institute for Physical Chemistry and Electrochemistry in Berlin.
127. This radio address was recorded during Einstein's visit to the Rádio Sociedade studios on 7 May 1925 and published in *O Jornal* on 8 May 1925. Published in *CPAE 2015*, Vol. 14, Appendix M, p. 999.
128. ALS (IsJNLI/Schwadron Mss. Collection, Einstein Collection). [AEA, 120 779]. Published in *Tolmasquim 2003*, p. 154, and in *CPAE 2015*, Vol. 14, Doc. 486, p. 773. Written on letterhead "Communidade Israelita Brasileira Direcção e Administração Do Gabinete do Presidente" and addressed "An die israelitische Gemeinde in Rio de Janeiro."
129. A reception had been held in Einstein's honor by the Jewish community at the Automobile Club of Brazil on 9 May. Einstein had visited both the Zionist Center and the Scholem Aleichem Library on 10 May (see text of diary, this volume, entries for 9 and 10 May 1925).
130. Isaiah Raffalovich.
131. ALS (NoONPPC). [AEA, 71 113]. Published in *CPAE 2015*, Vol. 14, Doc. 488, p. 777. Written on letterhead "Hamburg-Südamerikanische Dampfschifffahrts-Gesellschaft 'Cap Norte'" and addressed "An den Vorsitzenden des norwegischen Nobel-Komites."
132. The chairman of the committee was the Norwegian Law Professor Fredrik Stang (1867–1941).
133. Cândido Rondon.
134. During his second visit to Rio de Janeiro, Einstein met the professors of the Polytechnic School José Agostinho dos Reis, Arthur Getúlio das Neves, and Ignácio do Amaral. The film was shown to him on 11 May 1925 at the Brazilian Press Association (see text of diary, this volume).
135. ALSX. [AEA, 75 643]. Published in *CPAE 2015*, Vol. 14, Doc. 489, pp. 777–778. Written on letterhead "Hamburg-Südamerikanische Dampfschifffahrts-Gesellschaft 'Cap Norte.'"
136. He was actually returning on 31 May (see Text 26 in the Additional Texts section of this volume).
137. Antonio Pacheco Leão.
138. The sixth session of the International Committee on Intellectual Cooperation.
139. Hans Albert and Eduard Einstein.

140. ALS. [AEA, 37 394]. Published in *Lohmeier and Schell 2005*, pp. 201–202, and in *CPAE 2015*, Vol. 14, Doc. 490, p. 779. Written on letterhead "Hamburg-Südamerikanische Dampfschifffahrts-Gesellschaft 'Cap Norte.'" Glitscher (1886–1945) was a staff physicist at the factory of Hermann Anschütz-Kaempfe.

141. ALX. [AEA, 7 352]. Published in *Einstein and Besso 1972*, p. 204, and in *CPAE 2018*, Vol. 15, Doc. 2, pp. 40–41.

142. ALSX. [AEA, 75 642]. Published in *CPAE 2018*, Vol. 15, Doc. 7, pp. 47–48.

143. For the promise to bring the cacti back from Rio de Janeiro, see Text 30 in the Additional Texts in this volume. Margot Einstein.

144. Tete was Eduard Einstein's nickname. Einstein frequently vacationed with his sons at the house of his friend Hermann Anschütz-Kaempfe in Kiel.

145. On the acquisition of the butterfly collection, see Text 30 in the Additional Texts section of this volume.

146. ALS (Christie's online auction sale 16447, 2–9 May 2018, lot 36). [AEA, 97 163]. Published in *CPAE 2021*, Vol. 15, Doc. 18 in Vol. 16, pp. 26–27.

147. In this sentence, Einstein crossed out "South America" and wrote "Argentina" in its stead.

148. ALS (CaPsCA). [AEA, 17 357]. Published in *CPAE 2018*, Vol. 15, Doc. 20, p. 67. Millikan (1868–1953) was chairman of the Executive Council of the California Institute of Technology (Caltech) in Pasadena, Professor of Physics, and director of the Norman Bridge Laboratory of Physics.

149. Millikan had repeatedly invited Einstein to spend a winter term at Caltech, most recently in April (see Robert A. Millikan to Einstein, 2 April 1925 [*CPAE 2015*, Vol. 14, Doc. 468] and *CPAE 2015*, Vol. 14, Introduction, p. lii).

References

Newspapers and Periodicals

A Noite (Rio de Janeiro)
Berliner Tageblatt (Berlin)
Crítica (Buenos Aires)
Dos Idische Vochenblat (Rio de Janeiro)
Die Deutsche La Plata Zeitung (Buenos Aires)
Die Presse (Buenos Aires)
El Bien Público (Montevideo)
El Día (Montevideo)
El País (Montevideo)
El Sol (Montevideo)
Gazeta das Noticias (Rio de Janeiro)
Illustração Brasileira (Rio de Janeiro)
Ingeniería (Montevideo)
Jornal do Brasil (Rio de Janeiro)
La Época (Buenos Aires)
La Mañana (Montevideo)
La Nación (Buenos Aires)
La Prensa (Buenos Aires)
La Razón (Buenos Aires)
La Tribuna Popular (Montevideo)
Mundo Israelita (Buenos Aires)
New York Times (New York)
Nieuwe Rotterdamsche Courant (Rotterdam)
O Imparcial (Rio de Janeiro)
O Jornal (Rio de Janeiro)
O Paiz (Rio de Janeiro)
Revista Fon-Fon (Rio de Janeiro)
Vossische Zeitung (Berlin)

Books and Articles

Andrews 2010 Andrews, George Reid. *Blackness in the White Nation. A History of Afro-Uruguay*. Chapel Hill: University of North Carolina Press, 2010.

Asúa and Hurtado de Mendoza 2006 Asúa, Miguel de, and Diego Hurtado de Mendoza. *Imágenes de Einstein. Relatividad y cultura en el mundo y en la Argentina*. Buenos Aires: Eudeba, 2006.

Avni 1972 Avni, Haim. "Jewish Communities in Latin America." In *World Politics and the Jewish Condition: Task Force Studies Prepared for the American Jewish Committee on the World of the 1970s*, ed. Louis Henkin, pp. 256–274. New York: Quadrangle Books, 1972.

Bethell 1986 Bethell, Leslie. *The Cambridge History of Latin America*. Vol. 5. *c.1870 to 1930*. Cambridge: Cambridge University Press, 1986.

Bethell 2010 ——. "Brazil and 'Latin America.'" *Journal of Latin American Studies* 42, no. 3 (2010): 457–485.

Beretta Curi 2018 Beretta Curi, Alcides. "Inmigración alemana en Uruguay. Los inicios y temprano desarrollo de un establecimiento agropecuario modelo: Los Cerros de San Juan (1854–1929)." *RIVAR* 5, no. 13 (2018): 78–97.

Bernecker and Fischer 1992 Bernecker, Walther L. and Thomas Fischer. "Deutsche in Lateinamerika." In *Deutsche im Ausland. Fremde in Deutschland. Migration in Geschichte und Gegenwart*, ed. Klaus J. Bade, pp. 197–214. Munich: Beck, 1993.

Caffarelli 1979 Caffarelli, Roberto Vergara. "Einstein e o Brasil." *Ciência & Cultura* 31, no. 12 (1979): 1436–1455.

Calaprice et al. 2015 Calaprice, Alice, Daniel Kennefick, and Robert Schulmann, eds. *An Einstein Encyclopedia*. Princeton, NJ: Princeton University Press, 2015.

Chaucer 1924 Chaucer, Geoffrey. *Canterbury-Erzählungen. Mit 26 farb. Taf.* Nach Wilhelm Hertzbergs Übers. neu hrsg. v. John Koch. Berlin: Stubenrauch, 1924.

CPAE 1993 Einstein, Albert. *The Collected Papers of Albert Einstein*. Vol. 5, *The Swiss Years: Correspondence, 1902–1914*, ed. Martin J. Klein et al. Princeton, NJ: Princeton University Press, 1993.

CPAE 1996 ——. *The Collected Papers of Albert Einstein*. Vol. 6, *The Berlin Years: Writings, 1914–1917*, ed. Martin J. Klein et al. Princeton, NJ: Princeton University Press, 1996.

CPAE 1998 ——. *The Collected Papers of Albert Einstein*. Vol. 8, *The Berlin Years: Correspondence, 1914–1918*, ed. Robert Schulmann et al. Princeton, NJ: Princeton University Press, 1998.

CPAE 2004 ——. *The Collected Papers of Albert Einstein*. Vol. 9, *The Berlin Years: Correspondence, January 1919–April 1920*, ed. Diana Kormos Buchwald et al. Princeton, NJ: Princeton University Press, 2004.

CPAE 2006 ——. *The Collected Papers of Albert Einstein*. Vol. 10, *The Berlin Years: Correspondence, May–December 1920, and Supplementary Correspondence, 1909–1920*, ed. Diana Kormos Buchwald et al. Princeton, NJ: Princeton University Press, 2006.

CPAE 2009 ——. *The Collected Papers of Albert Einstein*. Vol. 12, *The Berlin Years: Correspondence, January–December 1921*, ed. Diana Kormos Buchwald et al. Princeton, NJ: Princeton University Press, 2009.

CPAE 2012 ——. *The Collected Papers of Albert Einstein*. Vol. 13, *The Berlin Years: Writings & Correspondence, January 1922–March 1923*, ed. Diana Kormos Buchwald et al. Princeton, NJ: Princeton University Press, 2012.

CPAE 2015 ——. *The Collected Papers of Albert Einstein*. Vol. 14, *The Berlin Years: Writings & Correspondence, April 1923–May 1925*, ed. Diana Kormos Buchwald et al. Princeton, NJ: Princeton University Press, 2015.

CPAE 2018 ——. *The Collected Papers of Albert Einstein*. Vol. 15, *The Berlin Years: Writings & Correspondence, June 1925–May 1927*, ed. Diana Kormos Buchwald et al. Princeton, NJ: Princeton University Press, 2018.

CPAE 2021 ——. *The Collected Papers of Albert Einstein*. Vol. 16, *The Berlin Years: Writings & Correspondence, June 1927–May 1929*, ed. Diana Kormos Buchwald et al. Princeton, NJ: Princeton University Press, 2021.

Einstein 1916 Einstein, Albert. "Die Grundlage der allgemeinen Relativitätstheorie." *Annalen der Physik* 49 (1916): 769–822.

Einstein 1917 ——. *Über die spezielle und die allgemeine Relativitätstheorie. (Gemeinverständlich.)* Braunschweig: Vieweg, 1917.

Einstein 1923a ———. "Zur allgemeinen Relativitätstheorie." *Preußische Akademie der Wissenschaften* (Berlin). *Physikalisch-mathematische Klasse. Sitzungsberichte* (1923): 32–38.
Einstein 1923b ———. "Bemerkung zu meiner Arbeit 'Zur allgemeinen Relativitätstheorie.'" *Preußische Akademie der Wissenschaften* (Berlin). *Physikalisch-mathematische Klasse. Sitzungsberichte* (1923): 76–77.
Einstein 1923c ———. "Zur affinen Feldtheorie." *Preußische Akademie der Wissenschaften* (Berlin). *Physikalisch-mathematische Klasse. Sitzungsberichte* (1923): 137–140.
Einstein 1924 ———. "Bietet die Feldtheorie Möglichkeiten für die Lösung des Quantenproblems?" *Preußische Akademie der Wissenschaften* (Berlin). *Physikalisch-mathematische Klasse. Sitzungsberichte* (1923): 359–364.
Einstein 1925 ———. "Einheitliche Feldtheorie von Gravitation und Elektrizität." *Preußische Akademie der Wissenschaften* (Berlin). *Physikalisch-mathematische Klasse. Sitzungsberichte* (1925): 414–419.
Einstein 2018 ———. *The Travel Diaries of Albert Einstein: The Far East, Palestine and Spain, 1922–1923*, ed. Ze'ev Rosenkranz. Princeton, NJ: Princeton University Press, 2018.
Eisinger 2011 Eisinger, Josef. *Einstein on the Road*. Amherst, NY: Prometheus Books, 2011.
Elkin 2014 Elkin, Judith Laikin. *The Jews of Latin America*. 3rd ed. Boulder: Lynne Rienner Publishers, 2014.
Fausto and Fausto 2014 Fausto, Boris, and Sergio Fausto. *A Concise History of Brazil*. 2nd ed. Cambridge: Cambridge University Press, 2014.
Fölsing 1997 Fölsing, Albrecht. *Albert Einstein: A Biography*. New York: Penguin Books, 1997.
Frank 1947 Frank, Philipp. *Einstein: His Life and Times*. New York: A. A. Knopf, 1947.
Gangui and Ortiz 2005 Gangui, Alejandro and Eduardo L. Ortiz. "Marzo–abril 1925: crónica de um mes agitado. Albert Einstein visita la Argentina." *Todo es Historia* 454 (2008): 22–30.
Gangui and Ortiz 2008 ———. "Einstein's Unpublished Opening Lecture for His Course on Relativity Theory in Argentina, 1925." *Science in Context* 21 (2008): 435–450.
Gangui and Ortiz 2014 ———. "Albert Einstein en la Argentina: en impacto científico de su visita." In *Visitas culturales en la Argentina, 1898–1936*, ed. Paula Bruno, pp. 167–190. Buenos Aires: Biblos, 2014.
Gaviola 1952 Gaviola, Enrique. "Alberto Einstein. Premio Nobel de Fisica, 1921." *Ciencia e Investigación* 8, no. 5 (1952): 234–238.
Gherman 2017 Gherman, Michel. "The Beginnings of Brazilian Zionism: Historical Formation and Political Developments." In *Jews and Jewish Identities in Latin America*, ed. Margalit Bejerano et al., pp. 190–207. Nashville, TN: Academic Studies Press, 2017.
Glick 1999 Glick, Thomas F. "Between Science and Zionism: Einstein in Brazil." *Episteme* 9 (1999): 101–120.
Goethe 1952 Goethe, Johann Wolfgang von. *Werke*. Vol. 1. Hamburg: Wegner, 1952.
Grundmann 2004 Grundmann, Siegfried. *Einsteins Akte. Wissenschaft und Politik—Einsteins Berliner Zeit*. 2nd ed. Berlin: Springer, 2004.
Hedges 2015 Hedges, Jill. *Argentina: A Modern History*. London: I. B. Tauris, 2015.

Horowitz 1962 Horowitz, Irving Louis. "The Jewish Community of Buenos Aires." *Jewish Social Studies* 24, no. 4 (1962): 195–222.
James 1897 James, William. *The Will to Believe and Other Essays in Popular Philosophy*. New York: Longmans, 1897.
James 1907 ———. *Pragmatism: A New Name for Some Old Ways of Thinking*. New York: Longmans, 1907.
Jerusalem 1928 Jerusalem, Else. *Steinigung in Sakya. Ein Schauspiel in drei Akten*. Berlin: Erich Reiss, 1928.
Kaplan 1997 Kaplan, E. Ann. *Looking for the Other: Feminism, Film, and the Imperial Gaze*. New York: Routledge, 1997.
Koigen 1925 Koigen, David. *Apokalyptische Reiter. Aufzeichnungen aus der jüngsten Geschichte*. Berlin: Erich Reiss, 1925.
König 1992 König, Hans-Joachim. "Das Lateinamerikabild in der deutschen Historiographie." In *Das Bild Lateinamerikas im deutschen Sprachraum. Ein Arbeitsgespräch an der Herzog August-Bibliothek, Wolfenbüttel, 15.–17. März 1989*, ed. Gustav Siebenmann and Hans-Joachim König, pp. 209–229. Tübingen: Max Niemayer Verlag, 1992.
Le Bon 1919 Le Bon, Gustave. *Aphorismes du temps présent*. Paris: E. Flammarion, 1919.
Leerssen 2000 Leerssen, Joep. "The Rhetoric of National Character. A Programmatic Survey." *Poetics Today* 21, no. 2 (2000): 267–292.
Lesser 1989 Lesser, Jeffrey Howard. "The Pawns of the Powerful: Jewish Immigration to Brazil, 1904–1945," PhD diss. (New York University, 1989).
Levine 1987 Levine, Robert M. "Adaptive Strategies of Jews in Latin America." In *The Jewish Presence in Latin America*, ed. Judith Laikin Elkin and Gilbert W. Merkx, pp. 71–84. Boston: Allen & Unwin, 1987.
Liebman 1973 Liebman, Seymour B. "Latin American Jews: Ethnicity and Nationalism." *Jewish Frontier* 40 (July–August 1973): 8–13.
Lohmeier and Schell 2005 Lohmeier, Dieter, and Bernhardt Schell, eds. *Einstein, Anschütz und der Kieler Kreiselkompaß. Der Briefwechsel zwischen Albert Einstein und Hermann Anschütz-Kaempfe und andere Dokumente*. 2nd ed. Kiel: Raytheon Marine GmbH, 2005.
Lynch 2012 Lynch, John. *New Worlds: A Religious History of Latin America*. New Haven: Yale University Press, 2012.
MacLachlan 2003 MacLachlan, Colin M. *A History of Modern Brazil: The Past Against the Future*. Wilmington, DE: Scholarly Resources Inc., 2003.
Meyerson 1925 Meyerson, Emile. *La déduction relativiste*. Paris: Payot, 1925.
Miles and Brown 2003 Miles, Robert, and Malcolm Brown. *Racism*, 2nd ed. London: Routledge, 2003.
Minguet 1992 Minguet, Charles. "Alexander Humboldt und die Erneuerung des Lateinamerikabildes." In *Das Bild Lateinamerikas im deutschen Sprachraum. Ein Arbeitsgespräch an der Herzog August-Bibliothek Wolfenbüttel, 15.–17. März 1989*, ed. Gustav Siebenmann and Hans-Joachim König, pp. 107–125. Tübingen: Max Niemeyer Verlag, 1992.
Mirelman 1990 Mirelman, Victor A. *Jewish Buenos Aires, 1890–1930. In Search of an Identity*. Detroit: Wayne State University Press, 1990.

Moraes 2019 Moraes, Diego. *Einstein en Uruguay:* Crónica de un viaje histórico. Montevideo: Ediciones B, 2019 (Kindle).

Moreira 1995 Moreira, Ildeu de Castro. "A Recepção das Idéias da Relatividade no Brasil." In *Einstein e o Brasil*, ed. Ildeu de Castro Moreira and Antonio Augusto Passos Videira, pp. 177–206. Rio de Janeiro: Editora da UFRJ, 1995.

Mudimbe-Boyi 1992 Mudimbe-Boyi, Elisabeth. "Travel, Representation, and Difference, or how can one be a Parisian?" *Research in African Literatures Journal* 23 (1992): 25–39.

Newton 1977 Newton, Ronald C. *German Buenos Aires, 1900–1933: Social Change and Cultural Crisis.* Austin: University of Texas Press, 1977.

Oberacker 1979 Oberacker, Karl-Heinz, Jr. "Die Deutschen in Brasilien." In *Die Deutschen in Lateinamerika*, ed. Hartmut Fröschle, pp. 169–300. Tübingen: Erdmann, 1979.

Ortiz 1995 Ortiz, Eduardo L. "A Convergence of Interests: Einstein's Visit to Argentina in 1925." *Ibero-Amerikanisches Archiv* 21, nos 1–2 (1995): 67–126.

Ortiz and Otero 2001 Ortiz, Eduardo L. and Mario H. Otero. "La visita de Einstein a Montevideo en 1925." *Mathesis* serie II (2001): 1–35.

Paty 1999 Paty, Michel. "La réception de la théorie de la relativité au Brésil et l'influence des traditions scientifiques européennes." *Archives internationales d'histoire des sciences* 49, no. 143 (1999): 331–368.

Raffalovich 1952 Raffalovich, Isaiah. *Tziunim ve'Tamrurim. Be shiv'im shnot nedudim, 5642–5712. Autobiografia.* Tel Aviv: Dfus Shoshani, 1952.

Renn 2013 Renn, Jürgen. "Einstein as a Missionary in Science." *Science & Education* 22 (2013): 2569–2591.

Romero 2013 Romero, Luis Alberto. *A History of Argentina in the Twentieth Century*. Updated and rev. ed. University Park, PA: Pennsylvania State University Press, 2013.

Rosenkranz 2011 Rosenkranz, Ze'ev. *Einstein before Israel: Zionist Icon or Iconoclast?* Princeton, NJ: Princeton University Press, 2011.

Rosenswaike 1960 Rosenswaike, Ira. "The Jewish Population of Argentina: Census and Estimate, 1887–1947." *Jewish Social Studies* 22, no. 4 (1960): 195–214.

Santos and Hallewell 2002 Santos, Sales Augusto dos, and Laurence Hallewell. "Historical Roots of the 'Whitening' of Brazil." *Latin American Perspectives* 29, no. 1 (2002): 61–82.

Sapolinsky 1963 Sapolinsky, Asher. "The Jewry of Uruguay." *In the Dispersion* 2 (1963): 74–88.

Sauer 2012 Sauer, Tilman. "On Einstein's Early Interpretation of the Cosmological Constant." *Annalen der Physik (Berlin)* 524 (2012): A135–A138.

Sayen 1985 Sayen, Jamie. *Einstein in America: The Scientist's Conscience in the Age of Hitler and Hiroshima.* New York: Crown Publishers, Inc., 1985.

Schenkolewski-Kroll 2017 Schenkolewski-Kroll, Silvia. "Informal Jewish Education: Argentina's Hebraica Society." In *Jews and Jewish Identities in Latin America*, ed. Margalit Bejerano et al., pp. 73–90. Nashville, TN: Academic Studies Press, 2017.

Schonebaum 1998 Schonebaum, Dieter. "Alemanes, Judíos y Judíos Alemanes en el Uruguay de los Años 1920 y 1930." *Jahrbuch für Geschichte Lateinamerikas* 35 (1998): 219–238.

Segura-Ubiergo 2007 Segura-Ubiergo, Alex. *The Political Economy of the Welfare State in Latin America*. Cambridge: Cambridge University Press, 2007.

Seth and Knox 2006 Seth, R., and C. Knox. *Weimar Germany between Two Worlds. The American and Russian Travels of Kisch, Toller, Holitscher, Goldschmidt, and Rundt*. New York: Peter Lang, 2006.

Shaw 1919 Shaw, G. Bernard. *Mensch und Übermensch. Eine Komödie und eine Philosophie*. Siegfried Trebitsch, trans. Berlin: S. Fischer, 1919.

Siebenmann 1988 Siebenmann, Gustav. "Das Lateinamerikabild der Deutschen. Quellen, Raster, Wandlungen." *Colloquium helveticum* 7 (1988): 57–82.

Siebenmann 1992a ——. "Methodisches zur Bildforschung." In *Das Bild Lateinamerikas im deutschen Sprachraum. Ein Arbeitsgespräch an der Herzog August-Bibliothek, Wolfenbüttel, 15.–17. März 1989*, ed. Gustav Siebenmann and Hans-Joachim König, pp. 1–17. Tübingen: Max Niemayer Verlag, 1992.

Siebenmann 1992b ——. "Das Lateinamerikabild in deutschsprachigen literarischen Texten." In *Das Bild Lateinamerikas im deutschen Sprachraum. Ein Arbeitsgespräch an der Herzog August-Bibliothek, Wolfenbüttel, 15.–17. März 1989*, ed. Gustav Siebenmann and Hans-Joachim König, pp. 181–207. Tübingen: Max Niemayer Verlag, 1992.

Silva da Silva 2005 Silva da Silva, Circe Mary. "The Theory of Relativity in Brazil: Reception, Opposition and Public Interest." In *Albert Einstein—Chief Engineer of the Universe: One Hundred Authors for Einstein*, ed. Jürgen Renn, pp. 294–297. Weinheim: Wiley–VCH, 2005.

Skidmore 1995 Skidmore, Thomas E. "Fact and Myth: Discovering a Racial Problem in Brazil." In *Population, Ethnicity, and Nation-Building*, ed. Calvin Goldscheider, pp. 91–117. New York: Routledge, 1995.

Spreitzer 2016 Spreitzer, Brigitte: "Else Jerusalem—eine Spurensuche." In *Else Jerusalem: Der heilige Skarabäus. Roman*, pp. 545–608. Vienna: DVB Verlag, 2016.

Stürzer 1993 Stürzer, Anne. *Dramatikerinnen und Zeitstücke. Ein vergessenes Kapitel der Theatergeschichte von der Weimarer Republik bis zur Nachkriegszeit.* Stuttgart: J. B. Metzler, 1993.

Summers 1999 Summers, William C. *Félix d'Herelle and the Origins of Molecular Biology*. New Haven and London: Yale University Press, 1999.

Tolmasquim 2003 Tolmasquim, Alfredo Tiomno. *Einstein. O viajante da relatividade na América do Sul*. Rio de Janeiro: Vieira & Lent, 2003.

Tolmasquim 2012 ——. "Science and Ideology in Einstein's Visit to South America in 1925." In *Einstein and the Changing Worldview of Physics*, ed. Christoph Lehner, Jürgen Renn, and Matthias Schemmel, pp. 117–133. New York: Springer, 2012.

Tolmasquim and Moreira 2002 Tolmasquim, Alfredo Tiomno and Ildeu de Castro Moreira, "Einstein in Brazil: The Communication to the Brazilian Academy of Science on the Constitution of Light." In *History of Modern Physics*, ed. Helge Kragh et al., pp. 229–242. Brussels: Brepols, 2002.

Vaz Ferreira 1909 Vaz Ferreira, Carlos. *El pragmatismo (exposición y crítica).* Montevideo: Tip. de la Escuela Nacional de Artes y Oficios, 1909.

Vaz Ferreira 1914 ——. *Le pragmatisme: Exposition et critique*. Montevideo: Barreiro & Ramos, 1914.

Walton 2009 Walton, John K. "Histories of Tourism." In *The SAGE Handbook of Tourism Studies*, ed. Tazim Jamal and Mike Robinson, pp. 115–129. London, SAGE, 2009.

Werner 1996 Werner, Harry. "Deutsche Institutionen und Schulen in Lateinamerika. Vielfalt und Wechselfälle des 19. und 20. Jahrhunderts." In *Deutsche in Lateinamerika—Lateinamerika in Deutschland*, ed. Karl Kohut, Dietrich Briesenmeister, and Gustav Siebenmann, pp. 182–196. Frankfurt am Main: Vervuert Verlag, 1996

World Bank 2002 The World Bank. *Higher Education in Brazil: Challenges and Options*. Washington, D.C.: World Bank, 2002.

Zapata 1979 Zapata, José A. Friedl. "Knobelbecher und Hakenkreuz gegen Gauchos und goldbraune Frauen. Das Lateinamerikabild in der deutschen und das Deutschlandbild in der lateinamerikanischen Literatur." *Zeitschrift für Kulturaustausch* 30 (1980): 50–59.

Internet Resources

"Ausstellung zu 160 Jahre diplomatische Beziehungen zu Uruguay." 15 July 2016. Accessed 3 May 2021. https://www.deutschland.de/en/node/3639.

Walther L. Bernecker. "Siedlungskolonien und Elitenwanderung. Deutsche in Lateinamerika: das 19. Jahrhundert." Accessed 3 May 2021. https://www.matices-magazin.de/archiv/15-deutsche-in-lateinamerika/deutsche-in-lateinamerika/.

Richard A. Campos. "Still Shrouded in Mystery: The Photon in 1925." Accessed 3 June 2021. https://www.yumpu.com/en/document/read/5228000/physics-0401044-pdf-arxiv

"Deutsch-uruguayische Beziehungen." Accessed 3 May 2021. https://www.pangloss.de/cms/index.php?page=uruguay.

Torsten Eßer. "Deutsche in Lateinamerika." Accessed 3 May 2021. http://www.torstenesser.de/download-text/Deutsche%20in%20Lateinamerika.pdf.

Eduardo L. Ortiz. "The emergence of theoretical physics in Argentina, Mathematics, mathematical physics and theoretical physics 1900–1950." Accessed 4 June 2021. *Proceedings of Science* (Héctor Rubinstein Memorial Symposium, 2010) 030. https://pos.sissa.it/109/030/pdf, pp. 1–17.

Index

Page numbers in *italics* refer to illustrations.
Albert Einstein is abbreviated to "AE" in subentries.